Quantum Computing and Communications

Springer
*London
Berlin
Heidelberg
New York
Barcelona
Hong Kong
Milan
Paris
Santa Clara
Singapore
Tokyo*

Michael Brooks (Ed.)

Quantum Computing and Communications

Springer

Michael Brooks, DPhil, BSc (Hons)

ISBN 1-85233-091-0 Springer-Verlag London Berlin Heidelberg

British Library Cataloguing in Publication Data
Quantum computing and communications
  1. Quantum computers 2. Telecommunication – Technological innovations
  I. Brooks, Michael
621.3'9'81
ISBN 1852330910

Library of Congress Cataloging-in-Publication Data
Brooks, Michael, 1970-
  Quantum computing and communications / Michael Brooks.
    p. cm.
  ISBN 1-85233-091-0 (alk. paper)
  1. Quantum computers. 2. Optical communications. I. Title.
QA76.889.B76 1999   99-19975
004.1–dc21   CIP

Apart from any fair dealing for the purposes of research or private study, or criticism or review, as permitted under the Copyright, Designs and Patents Act 1988, this publication may only be reproduced, stored or transmitted, in any form or by any means, with the prior permission in writing of the publishers, or in the case of reprographic reproduction in accordance with the terms of licences issued by the Copyright Licensing Agency. Enquiries concerning reproduction outside those terms should be sent to the publishers.

© Springer-Verlag London Limited 1999
Printed in Great Britain

The use of registered names, trademarks etc. in this publication does not imply, even in the absence of a specific statement, that such names are exempt from the relevant laws and regulations and therefore free for general use.

The publisher makes no representation, express or implied, with regard to the accuracy of the information contained in this book and cannot accept any legal responsibility or liability for any errors or omissions that may be made.

Typesetting: Camera ready by editor
Printed and bound by the Athenæum Press Ltd., Gateshead, Tyne & Wear
34/3830-543210 Printed on acid-free paper

Preface

We have, in the last few years, radically improved our grasp of the quantum world. Not just intellectually, either: our ability to manipulate real quantum systems has grown in equal measure with our understanding of their fundamental behavior. These two shoots - the intellectual and the practical harnessing of the quantum world - have sprung up at a time when a third shoot - information processing - has also been experiencing explosive growth. These three shoots are now becoming intertwined. Twisted together, our understanding of information processing, quantum theory and practical quantum control make for a strong new growth with enormous potential.

One must always be careful about using the word 'revolutionary' too readily. It is, however, difficult to find another word to describe the developments that have been taking place during the second half of the 1990s. In 1986 Richard Feynman, the visionary professor of physics, made a very interesting remark:

> "...we are going to be even more ridiculous later and consider bits written on one atom instead of the present 1011 atoms. Such nonsense is very entertaining to professors like me."

It is exceptionally unfortunate that Feynman did not live to see this 'nonsense' fully transformed into reality. He, more than anybody, would enjoy the fact that it is now possible to write information onto an atom, or indeed an ion or a photon. Furthermore, theorists and experimentalists have shown that this information can be processed and transmitted in ways that allow a seemingly absurd degree of power and control over the information. It is now possible to use one quantum particle to influence another particle that it has never met. It is possible to transmit information encoded in a single photon through the air and, on detection, to verify whether that information has been read by anyone else. Experiments are just beginning to string together quantum bits of information that promise massively parallel computing power, far beyond anything that classical machines can manage. In short, we are on the verge of the quantum information revolution.

There could have been no better time for the European Commission to fund a Pathfinder Project in Quantum Computing and Communications* to look into this subject. This Project, which was carried out with the financial support of the Commission, within the frame of the Esprit LTR Working Group 27126 QCEPP, facilitated the gathering and organization of a large amount of useful material about the field; I have freely and extensively used this material in this book. It is reproduced here with the kind permission of the Commission, but it does not necessarily reflect the views of the Commission.

It has been a great privilege for me to spend the last year being so closely involved with the Pathfinder Project. Some of the members of the Project's Working Group have laid the theoretical or experimental foundations in important areas of this field; I have taken great pleasure in working alongside them, and am grateful for all their assistance in drafting parts of this volume. As a journalist I often have to skim the surface of a subject, taking in its essence, but with little time to consider its implications or to examine its fundamental basis. My involvement with the Pathfinder Project has enabled me to investigate this fascinating area to my heart's content, often learning directly from those who have originated the concepts. I am grateful to the European Commission, and to all of the Working Group for their input and advice. However, I take full responsibility for any errors or omissions that have crept into this volume.

Finally, I would like to acknowledge the important role played by Brian Oakley, Chairman of the Pathfinder Project, and Charles Ross, its Honorary Secretary. Without their seemingly boundless energy and enthusiasm, it is likely that little of this material would ever have been gathered.

Michael Brooks, January 1999

---

* The term 'Quantum Information Processing' (QIP) is also used to describe Quantum Computing and Communications throughout this book.

Contents

Section I	A Wide Perspective	
Chapter 1	**Introduction**	
	Michael Brooks	3
	1.1 Exploiting the Quantum World	3
	1.2 Historical Background	5
	1.3 Worldwide Efforts in QCC	6
Chapter 2	**The Fundamentals of Quantum Information**	
	Michael Brooks	9
Chapter 3	**Quantum Computer Science**	
	Michael Brooks	17
	3.1 Introduction	17
	3.2 Algorithms and the Complexity Problem	18
	3.3 The Quantum Computation Answer	19
	3.4 Quantum Algorithms	20
	3.5 Quantum Logic Gates and Networks	21
	3.6 Obstacles	23
	3.7 A Workable Solution: Quantum Error Correction	23
	3.8 Conclusions	24
Chapter 4	**Experimental Realizations**	
	Michael Brooks	27
	4.1 Introduction	27
	4.2 Trapped Ions	27
	4.3 Nuclear Magnetic Resonance	29
	4.4 Cavity Quantum Electrodynamics	30
	4.5 Quantum Dots	31
Chapter 5	**Optical Technologies for Quantum Computing and Communications**	
	Michael Brooks	33
	5.1 Introduction	33
	5.2 Using the Quantum Nature of Light	34
	5.2.1 Potential	34
	5.2.2 Problems	34

	5.3 Quantum Noise in Optical Communications	35
	5.4 Generic Technologies in Quantum Communications	35
	5.4.1 Nonlinear Optics	36
	5.4.2 Cavity Quantum Electrodynamics	36
	5.5 Operations Performed on Optical Signals	37
	5.5.1 Signal Generation	37
	5.5.2 Detection	38
	5.5.3 Attenuation	38
	5.5.4 Distribution	39
	5.5.5 Amplification	39
	5.6 Conclusions: Towards the Second Generation	40
Chapter 6	**Applications** *Michael Brooks*	43
	6.1 Introduction	43
	6.2 Emerging Technologies	44
	6.2.1 Quantum Cryptography	44
	6.2.2 Quantum Repeaters	44
	6.2.3 Quantum Simulators	45
	6.2.4 Metrology and Few Photon Applications	46
	6.3 Conclusions: Measuring Progress	47
Chapter 7	**A Note on the Question of Scaling: Decoherence and Error Correction** *Michael Brooks*	49
Section II	**Personal Perspectives**	
Chapter 8	**Solid State Quantum Computation: Prospects, Proposals, and Prejudices** *Bruce Kane, University of New South Wales, Sydney, Australia*	53
Chapter 9	**Information is Physical, But Slippery** *Rolf Landauer, IBM, Yorktown Heights*	59
Chapter 10	**Nanocircuitry, Defect Tolerance and Quantum Computing: Architectural and Manufacturing Considerations** *R. Stanley Williams, Quantum Structures Research Initiative, Hewlett-Packard Laboratories*	63
Chapter 11	**Quantum Computing and NMR** *Jonathan A. Jones, University of Oxford*	71
Chapter 12	**Quantum Networks and Quantum Algorithms** *Vlatko Vedral, University of Oxford*	79

Chapter 13	**Quantum Cryptography** *Richard Hughes, Physics Division, Los Alamos National Laboratory* 87	
Section III	**A Perspective for the Future**	
Chapter 14	**Realizing the Potential of Quantum Information Processing** *Michael Brooks* 97	
	14.1 Prospects for Quantum Computing	97
	14.2 Prospects for Special Applications	98
	14.2.1 Quantum Simulation	98
	14.2.2 Limited Qubit and Reduced-Noise High Precision Applications	98
	14.2.3 Secure Communications	99
	14.3 Meeting The Needs of the QIP Field	99
	14.3.1 The Need for Academic Focus	100
	14.3.2 The Need for Industrial Focus	100
	14.3.3 The Need for Awareness	101
Chapter 15	**The Role of Europe** *Michael Brooks* 103	
	15.1 The Pioneering Stage	103
	15.2 Today	104
	15.2.1 Multi-Disciplinary or Trans-Disciplinary Nature of the Community	104
	15.2.2 The Numbers Involved	104
	15.2.3 The Geographical Spread	105
	15.2.4 The Industrial Scene	105
	15.2.5 Summary of the European Scene	106
Chapter 16	**Quantum Computing and Communications: A View from the USA** *Colin P. Williams, Jet Propulsion Laboratory, California Institute of Technology* 107	
	16.1 Introduction	107
	16.2 What Works Well	108
	16.3 What Does Not Work So Well	108
	16.4 NASA/JPL	109
	16.5 Lessons Learned from Experience with NASA	111
	16.6 Opportunities for Europe	111
	16.7 Commercialization	112
	16.8 Recommendations	113
	16.8.1 Programmatic Recommendations	113
	16.8.2 Technology Solutions	113
	16.8.3 Quantum Computer Science	113
	16.8.4 Education and Training	114

| Section IV | Reference materials |

| Chapter 17 | **Quantum Information Processing:**
A Brief Overview of Recent Advances |

Antonella Karlson, StarLab, Riverland nv 119
- 17.1 Introduction 119
- 17.2 The Underlying Physical System 120
- 17.2.1 Quantum Bits and Quantum Superpositions 120
- 17.2.2 Quantum Gates 120
- 17.2.3 Quantum Parallelism 121
- 17.3 Fundamentals of Quantum Information 121
- 17.3.1 Entanglement 121
- 17.3.2 Quantum Dense Coding 122
- 17.3.3 Quantum Teleportation 123
- 17.4 Quantum Cryptography 123
- 17.4.1 Standard Cryptosystems 123
- 17.4.2 Quantum Key Distribution 124
- 17.5 Quantum Computing 125
- 17.5.1 Quantum Algorithms 125
- 17.5.2 Grover's Search Algorithm 126
- 17.5.3 Period Finding and Shor's Factorization Algorithm 127
- 17.5.4 Minimum Requirements for any Quantum System to Be a Quantum Computer 128
- 17.5.5 Simulation of Other Quantum Systems 128
- 17.6 Quantum Decoherence 129
- 17.6.1 What is Decoherence? 129
- 17.6.2 Quantum Error Correction 130
- 17.7 Experimental Realizations 131
- 17.7.1 Systems with Few Degrees of Freedom 131
- 17.7.2 Macroscopic Systems 132
- 17.8 Conclusions 132

| Chapter 18 | **Categories and Definitions** .... 137 |

- 18.1 Definitions and Glossary of Terms 137
- 18.1.1 Definition of Quantum Computing and Communications 137
- 18.1.2 Sub-Sector Definitions 138
- 18.1.3 Some Relevant Terms 140
- 18.2 The Disciplines that Contribute to Quantum Information Processing 142

| Appendix A | **The Pathfinder Project** .... 145 |

- A.1 Pathfinder Activities 145
- A.1.1 Newsletters 145
- A.1.2 Database of Active Individuals and Organizations 146
- A.1.3 Taxonomy of the Subject 146
- A.1.4 Helsinki Conference 146

	A.1.5 QIP Report	147
	A.2 Pathfinder Project Partners	147
Appendix B	**A Preliminary QIP Roadmap**	149
Index		151

Section I

A Wide Perspective

An examination of the whole field of Quantum Computing and Communications requires some decisions to be taken about where to draw divisions between the various subject areas; the placement of these divisions is a subject that could sustain infinite debate. For the purposes of this volume, the field has been split into four main fields, each covered by a chapter in this section. These four chapters are preceded by an introduction to the subject of quantum information, and are followed by a note on the problems of decoherence.

Chapter 1
Introduction

Michael Brooks

1.1 Exploiting the Quantum World

Civilization has always advanced as people discovered new ways of exploiting various physical resources, such as materials, forces and energies. In the twentieth century, information was added to the list when the invention of computers allowed complex information processing to be performed outside human brains. The history of information technology has involved a sequence of changes from one type of physical realization to another; from gears to relays to valves to transistors to integrated circuits and so on. Now, developments of quantum information processing have reached the stage where one bit of information can be encoded in quantum systems, for example using two different polarizations of light, or two different electronic states of an atom. Matter on this scale obeys the laws of quantum mechanics, which are quite different from the classical laws followed by 'conventional' technologies. If an atom is used as a physical bit then quantum mechanics shows that apart from the two distinct electronic states the atom can be also prepared in a coherent superposition of the two states. This means that the atom is both in state 0 *and* state 1. In general, a quantum two-state system, called a quantum bit or a qubit, can be prepared in a superposition of its two logical states 0 and 1. Thus one qubit can encode at a given moment of time both 0 and 1. Strings of qubits in superposition states can be 'entangled' together to simultaneously encode, in principle at least, vast amounts of information.

Entanglement is one of the distinct properties of quantum systems (together with quantum superposition and probabilistic measurement among others) that makes quantum information processing so different from classical information technology. This phenomenon refers to the joint state of two or more quantum systems and describes correlations between them that are much stronger than any classical correlations. Entangled states offer the possibility to encode information in a completely new way. Let us assume we have two qubits and we want to encode two bits of information. The straightforward approach is to encode one bit of information onto each qubit separately. But using entangled states it is possible to do it in such a way that neither of the two qubits carries any well defined information on its own: all the information is encoded in their joint properties. Entanglement makes possible quantum teleportation, quantum error correction, quantum dense coding, etc. It is closely linked to the issue of non-locality in quantum theory.

This means that quantum technology, potentially, can offer much more than cramming more and more bits on to silicon and multiplying the clock-speed of microprocessors. It can support entirely new kinds of computation with qualitatively new algorithms based on quantum principles. It also offers very significant improvements - in speed, security and quality - in the technologies of information transfer.

Considering quantum computation first, it has been shown that quantum effects could allow the creation of a register of qubits. Such a register composed of three qubits, for example, can simultaneously represent the numbers from binary 000 to binary 111 - i.e. the decimal numbers 0 to 7. Computations performed on these entangled qubits thus have the potential to process simultaneously, offering new and exciting possibilities in information processing.

Algorithms have now been developed which show how quantum computation could lead to enormous advantages over classical computation, accomplishing tasks that are impossible, or impossibly time-consuming, to any classical computer - no matter what its clock speed or processing power. That list of tasks includes code-breaking factorization operations, which makes the development of quantum computation a matter of great importance to any institutions involved with matters of national or financial security. Practical implementation of these ideas is not without its very considerable difficulties and there can be no certainty that it will ever prove possible to build a useful quantum computer. But solutions to existing problems are continually being discovered and analyzed. Whatever the end result, the journey towards the implementation of quantum computation is already yielding valuable spin-off technologies and applications in fields such as communication security.

Entangled qubits are also at the heart of quantum communications, a field whose development is of fundamental interest to financial institutions and national security agencies. Research has demonstrated that measuring the properties of one of a pair of entangled photons would lead to an instantaneous change in the properties of the other half of the entangled pair, however far it was from the measured photon. Development of this work has taken entirely secure communications, using information encoded in entangled photon pairs, to a near-marketable development stage. The use of quantum effects are enabling new and extremely useful techniques to be discovered and implemented.

1.2 Historical Background

The subject of Quantum Information Processing is not new. In one sense the work stems from the recognition of the quantum nature of radiation by Max Planck in 1900, and the equation derived by Erwin Schrödinger in 1926 which provided a mathematical basis for quantum mechanics. This showed that if a quantum system was to operate as a computer it had to be capable of operating reversibly. The history of twentieth century computing, stemming from Turing's work in 1936, is based on a model of a computer that is not reversible. The super-computer of today is essentially of the same nature of machine as the early computers of the 1940s, as is the PC; they differ in memory and operating speed but not in fundamental operations. It has been recognized for many years that they have some practical limitations; for example in not being able to generate a truly random set of numbers; and not being able to perform certain calculations such as those that can be solved for smaller numbers, but at some point become too lengthy for any current or conceivable computer operating on the current classical principals. As the semiconductor switches, on which modern computers are based, have become smaller and smaller, it has become apparent that at some stage, probably within the next 10 to 20 years, the number of atoms in the structures will become so small that the classical laws will have to be replaced by the laws of quantum physics if their behavior is to be understood and predicted.

Perhaps the key breakthrough in quantum computation came in 1973 when Charles Bennett (IBM, Yorktown Heights) showed that a reversible Turing machine was a theoretical possibility. Then, in 1980, Paul Benioff (at that time at Argonne National Laboratory) formulated a reversible Turing machine, which could read, write and shift using quantum mechanical interactions. In 1982, Richard Feynman suggested that a quantum computer could simulate a quantum system efficiently, in a way that no classical computer could. And then, in 1985, David Deutsch (Oxford University) described how the quantum Turing machine might be built, in principle, and how the 'superposition' of 0s and 1s simultaneously led to quantum parallelism. During the 1980s, work was developing on ways of constructing the necessary quantum gates, and then experimental work on a variety of approaches to handling a limited number of quantum bits started in the 1990s. The problem of decoherence seemed to create a practical limitation on the use of quantum information processing, but work at various centers in the 1990s has shown how this might be overcome by error correction techniques. A very considerable boost to the practical interest in the subject came in 1994 when Peter Shor (then at Bell Labs) demonstrated an algorithm which showed how the superpolynomial time process for factorizing a large number on a classical computer could be reduced to an efficient polynomial time process on a quantum computer, a result of considerable interest to the cryptography community.

And amongst the algorithms for quantum computers that followed this breakthrough, Lov Grover developed in 1996 an algorithm that would reduce the time required to find a single item in an unsorted list in the square root of the time it would take on a classical computer. This work on algorithms has served to open up interest in the information processing community. And the fact that elementary versions of the algorithms can now be demonstrated (for example using the NMR approach) has triggered the coming together of the research workers in the quantum physics and IT communities.

The field of quantum communications had, like the ideas behind quantum computing, an early start. In 1935, Albert Einstein, Boris Podolsky and Nathan Rosen published a paper on the paradox that seemed to stem from quantum mechanics; measuring the properties of one of a pair of entangled photons in one place would lead to an instantaneous change in the properties of the other entangled particle, however far it was distant from the measured photon. This seemed so implausible to them that it was seen as an argument that some fallacy lay in the basic concept of quantum mechanics. But the effect has been demonstrated in practice in recent years by practical work both in Europe and the USA. This work has been used as the basis for a secure communication demonstration by DERA, BT, Geneva University, Los Alamos and others over tens of kilometers of installed fibber optics networks, in particular under Lake Geneva. While it is true that the practical system implementation means that the information is not obtainable faster than at the velocity of light, the process both enables a cryptographer's key to be transmitted absolutely securely and it ensures that any eavesdropping is detectable because to succeed it must destroy the unique entanglement link.

1.3 Worldwide Efforts in QCC

The subject of quantum information processing is one of the most interesting, stimulating, and potentially important of the new technologies that have opened up for many years. Though the fundamentals were known for many years, it is only in the last decade or so that the recognition of the potential processes that could achieve remarkable results have become clear. Making use of the quantum mechanics phenomenon of entanglement and superposition opens the way for natural parallelism in computing and for shorter term applications to secure communications, the simulation of quantum processes, for example, of molecular interactions, precision navigation, etc. While it is becoming increasingly clear how the building of a quantum computer could be achieved, it must remain in doubt if a powerful quantum computer can be built, and this is clearly some, if not many years away. But there is now no doubt that applications to secure communications will start in-service trials within the next few years.

There are various experimental techniques now under development; ion traps, cavity QED, quantum dots, quantum photon interferometry, and NMR, etc. The state of progress is changing fast but most techniques are currently only capable of handling one or two 'qubits'. However, it is expected that this will develop steadily with potential for ion traps, for example, to produce 10 or more qubits within the next 10 years. NMR has developed very fast in the last year, now

Introduction 7

being capable of handling three qubits, and will probably seem the leading technology over the next few years. But it is expected that the potential of this approach will run out once it attempts to cope with more than, say, six qubits. However, the technique is very important if only because it is already possible to demonstrate the quantum algorithms working, albeit with only a few qubits.

Though the fundamentals of the subject were first enunciated in Europe, work in the USA has played an important part in the opening up of the field. But Europe has always played a major part in most aspects of the subjects. In recent years, many of the most notable experimental results have been coming from laboratories in Europe, for example from the teams in Innsbruck, Geneva and Oxford, but with important results originating in many other centers round Europe as well. The work in Europe is widely dispersed except for a few major centers. But the numbers are growing steadily and new disciplines are being drawn in as the realization spreads of the intellectual excitement of the subject and of the potential applications.

Active industrial interest in the subject in Europe is currently largely restricted to those firms in the telecommunication industries where the potential of the quantum approach for distributing encryption keys and for generating inherently secure communications is widely recognized. There are a few small firms who are making an investment in the subject. The small number of interested firms is likely to change rapidly when it becomes widely known that commercial customers, for example in the financial institutions, are investing in the secure communications technology.

The subject is potentially important, and yet at the same time, the active research community is thinly and widely spread. It is growing and becoming very much a multi-disciplinary subject where physicists, both theoretical and experimental, are being joined by computer scientists, mathematicians, communications engineers, chemists, and others.

Chapter 2
The Fundamentals of Quantum Information

Michael Brooks

Ten, maybe fifteen years ago, the term 'Information Technology' was not one that could be used without explanation. Most people, certainly in the developed world, knew what computers were, and what they could be used for. But that was only in terms of 'end user' applications, like word processing, performing mathematical calculations or creating a database. The fact that a computer simply stores and processes information is not a concept that appears to stick easily in the mind.

Nowadays it seems that the persistent, widespread use of the term 'IT' in the modern world has made it synonymous with computing and communications technology. These 'technologies' are now found in almost every corner of the globe, and yet the 'information' part of the term remains somewhat ignored. Somehow, that is not surprising: the nature and the importance of information is only beginning to be understood in scientific circles. And just as this is taking place, a new paradigm - possibly the most important new concept for technological developments in the next century - is being explored. That new paradigm is quantum information.

It is perhaps misleading to call it new: since its inception in the 1920s, quantum theory has given rise to descriptions of systems that carry information. What is new is that research has unearthed the potential for novel means of processing information using the nature of quantum information.

One of the most important new insights into the field of information is that information itself is physical. It is not independent of the physical laws that govern whatever system is used to store and process it. So our new-found ability to encode information at the quantum scale - in particles that obey the laws of quantum mechanics rather than the 'classical' laws we encounter in our everyday lives - should open up entirely new behaviors in the information itself.

Grasping this potential, and developing it to the point of application, could revolutionize many aspects of modern data handling. If that sounds rather abstract (caution is a necessity when describing the potential of an undeveloped technology), then consider the possibilities - and they are just possibilities - afforded by a quantum computer that might take just a few minutes to break all the secret codes used by the world's security agencies. Or the potential for totally secure communication by encoding information in photons of light. Or the so-called 'teleportation' of quantum states of matter across large distances.

Information, fundamentally, consists of a bit: something that can take on one of two values. Traditionally, bits have been labeled as '0' or '1', but the label could be anything that distinguishes those values. The values of the bit can be encoded by a certain voltage level in a capacitor or by the alignment of particles on a magnetic tape, for example. The important point is that the two parameters - the '0' and '1', or 'on' and 'off'- are well separated in the parameter space in which they have been encoded. That allows the bit to be well defined, so that the information is reliable.

The high desirability of separate, well-defined states in classical information processing takes on a new dimension when we come to quantum systems, however. The characteristic that makes quantum information processing so radically different from classical methods is that the information can be in an undefined state and still be extremely useful - more so than when well-defined, in fact.

This is because of the phenomenon of quantum superposition. In a superposition of quantum states, the bit can be in both the '0' and '1' states at the same time. This feature of the quantum bit, or 'qubit', has no parallel in classical systems.

To make the concept more concrete, imagine an atom, whose electrons can exist at various discrete energy levels. We could label the lowest energy (ground) state as '0' and the next highest energy level as '1'. Information could then be encoded in the atom (just as it is in capacitors in classical computers) by controlling the energy state of the electron. But the atom can be made to enter a superposition of states, where the electron is in the state '0' and the state '1' simultaneously. Dealing with binary data, an atom can represent 0 or 1, or it can represent '0' *and* '1'.

This does require, however, that the atom remains 'unobserved': the act of observation, or measurement, implies obtaining a knowledge of the system which is simply forbidden. When measured, a superposition will 'collapse' into a randomly chosen single state. It is worth re-stating the randomness of this process: there is nothing to dictate which state the superposition will collapse into. There are many philosophical interpretations of this phenomenon (which cannot be discussed here), but there is no doubt that quantum systems in superposition states are extraordinary in their behavior. They are also extremely fragile, and must be 'handled' with the utmost care.

Assuming that our initial atom remains, undisturbed, in its superposition, we can place another atom next to it to give us two qubits. These two qubits can

The Quantum Mechanical Bit: Qubit

Two States:

$|0\rangle$ or $|1\rangle$

Quantum Mechanical Superposition:

e.g. $|\psi\rangle = \frac{1}{\sqrt{2}}|0\rangle + \frac{1}{\sqrt{2}}|1\rangle$

In this scheme, one qubit is in a superposition of the states zero and one. The number written in front of the states can be any complex number, as long as the absolute adds up to one. In the case above, the number indicates that, if a measurement is made, the result will be zero or one with equal probability (50%).

If we consider a pair of two-state particles (labeled with subscripts 1 and 2), there are four possibilities for the 'entangled' superposition state:

$|\psi^+\rangle = (\frac{1}{\sqrt{2}})(|0\rangle_1|1\rangle_2 + |1\rangle_1|0\rangle_2)$

$|\psi^-\rangle = (\frac{1}{\sqrt{2}})(|0\rangle_1|1\rangle_2 - |1\rangle_1|0\rangle_2)$

$|\phi^+\rangle = (\frac{1}{\sqrt{2}})(|0\rangle_1|0\rangle_2 - |1\rangle_1|1\rangle_2)$

$|\phi^-\rangle = (\frac{1}{\sqrt{2}})(|0\rangle_1|0\rangle_2 - |1\rangle_1|1\rangle_2)$

These states define the possibilities. In the first situation ($|\psi^+\rangle$), for example, the only defined matter is that the two qubits are in different states: there is no way of knowing which qubit is in which state. These entangled states are known as Bell states.

Figure 2.1: notations and basic concepts for superposition and entanglement in quantum information theory.

be 'entangled' such that their superpositions are linked; the two atoms therefore have the potential to represent 0, 1, 2 or 3.

An entangled pair of quantum objects store *all* the information about their states in their *joint* properties. To emphasize that point: there is *no* information in their *individual* properties. It is worth remarking that an entangled pair is not like a set of identical twins: in this case, each one of a twin carries information that is identical to the information the other twin carries. In entanglement neither of the particles, considered on its own, carries any information. This is what Erwin Schrödinger called 'the essence of quantum mechanics'. In principle, this entanglement phenomenon enables any number of qubits to be strung together in order to encode large numbers: three qubits can simultaneously encode binary '0' to binary '7', four qubits can encode binary '0' to binary '15'; entangle 250 atoms together and one could simultaneously encode more numbers than there are atoms in the known universe.

The prospect offered by quantum computing involves using an entangled state as the input to a computation. That would mean that the quantum computer could process all the possibilities at once. What comes out, of course, is also an entanglement - of the many possible answers generated by executing a function on the many entangled inputs. Extracting the desired information is not a trivial operation, but the algorithms that have so far been developed to display the usefulness of quantum information processing have shown that the exploitation of interference effects (where some information about the global system is contained within each of its constituent parts) can make the required information stand out like a melody from harmony.

Although there are grand theoretical schemes in place for quantum computation, in practice things are still somewhat more modest. It is fairly straightforward now to entangle two photons, and four nuclei have been entangled in a liquid, but moving beyond this remarkable achievement is proving difficult.

The practical implementation of two particle entanglement has, however, validated the fact that quantum mechanics changes the way information can be handled. The Bell states set out in Figure 2.1, for example, show that two entangled qubits can be put into any one of four different states by performing an operation on any one of the qubits. These operations could include, for example, 'flipping' the qubit (changing it from '0' to '1' or vice-versa) or adding a phase shift so that it changes sign. And no matter what their physical separation, an operation performed on one half of the pair will alter the other half. So two people who want to exchange information over some distance (they are almost always referred to as 'Alice' and 'Bob' by those in the quantum information community) can each have one of an entangled pair of particles. If Bob performs an operation on his particle according to the information he wishes to transmit, and then sends the photon to Alice, she will be able to infer the information from the change in her particle and the state of Bob's transmitted photon. Using this scheme, known as quantum dense coding, it is possible to send two bits of information using just one two-state photon - impossible to achieve using classical means.

So, although the difficulty of multi-particle entanglement may slow the progress of many-qubit quantum computing operations, it does not affect the

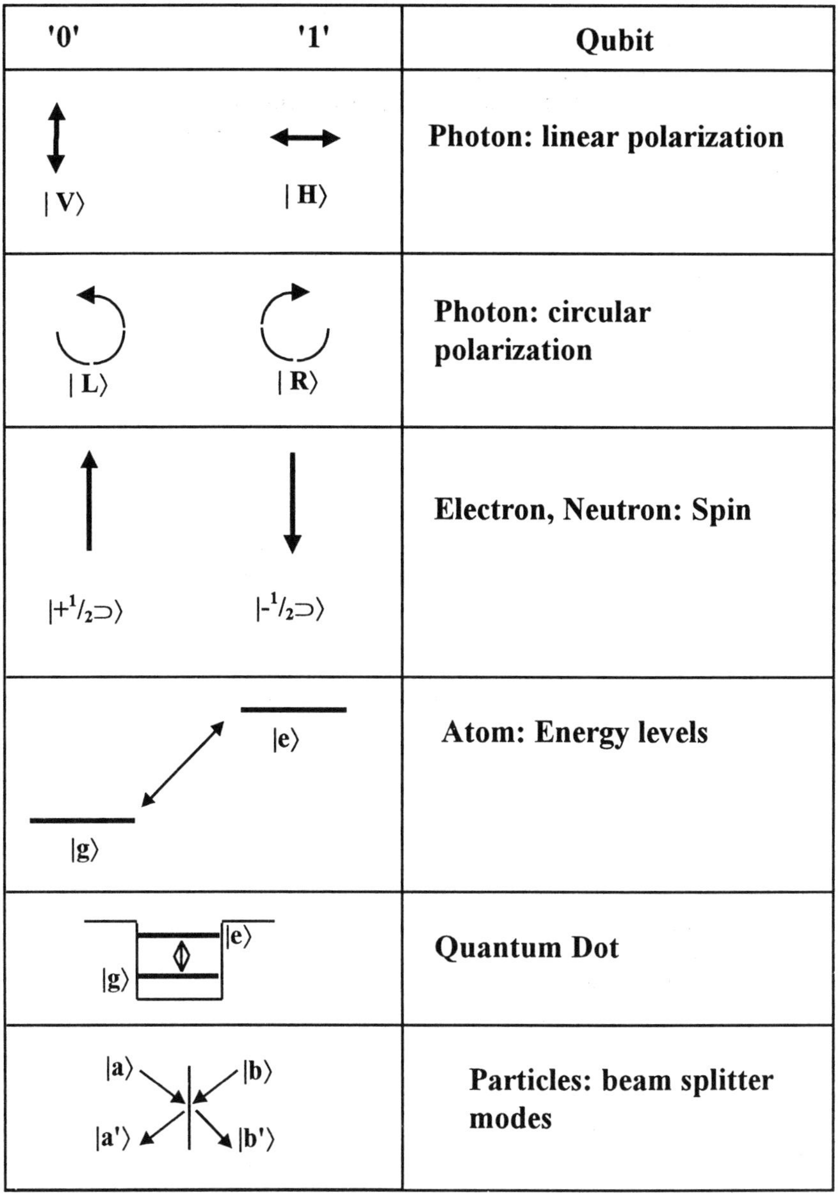

'0'	'1'	Qubit
$\|V\rangle$	$\|H\rangle$	**Photon: linear polarization**
$\|L\rangle$	$\|R\rangle$	**Photon: circular polarization**
$\|+^1/_2\supset\rangle$	$\|-^1/_2\supset\rangle$	**Electron, Neutron: Spin**
$\|g\rangle$	$\|e\rangle$	**Atom: Energy levels**
$\|g\rangle$	$\|e\rangle$	**Quantum Dot**
$\|a\rangle, \|a'\rangle$	$\|b\rangle, \|b'\rangle$	**Particles: beam splitter modes**

Figure 2.2: physical realizations of qubits

development of many other aspects of the quantum information field. The realization of quantum dense coding has led to quantum state teleportation, where the quantum state of a photon is recreated on another, distant photon.

Also achieved in experiments is 'entanglement swapping', where two pairs of entangled particles are used to establish an entanglement between two particles that have never met (Figure 2.3).

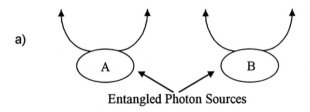

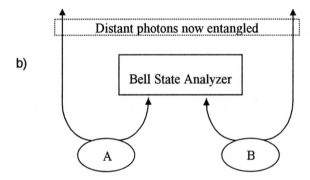

Figure 2.3: a) two sources each emit a pair of entangled photons. b) a Bell state measurement is carried out on one of each pair. The other two photons, which have never met, are now themselves entangled.

Performing a Bell measurement on one half of Pair A and one half of Pair B transfers the entanglement to the other halves of the pair. In this scheme, one ends up with the ability to affect the quantum state of a particle by performing a measurement on another particle that has never had any contact with the first. It is an astonishing and puzzling - but nonetheless experimentally proven - feature that demonstrates the new potential for communication technologies that is opened up in the quantum world.

The modern ability to manipulate the quantum mechanical properties of particles has enabled various technological revolutions to take place already. Quantum cryptography, for example, uses the fact that measurement affects a quantum system to create an entirely secure method of data transmission.

Information is encoded in the polarization states of photons and transmitted - through optical fibers or free space - over long distances. An eavesdropper disturbs the quantum state of the photon and so can always be detected. As discussed elsewhere in this volume, this technology is now nearing commercial viability.

Chapter 3
Quantum Computer Science

Michael Brooks

3.1 Introduction

Compare the Babbage differential engine with a modern computer. The technological gap is obvious and yet there is essentially no difference between what the two machines can do - one simply does things faster than the other. The history of computer technology has involved a sequence of changes from one type of physical realization to another. Moving from gears to relays to valves to transistors to integrated circuits has gradually increased the computing speed, but the capability of the machines has not changed.

Quantum technology can offer much more than cramming more and more bits onto silicon and multiplying the clock-speed of microprocessors. Quantum computers can be programmed in a qualitatively new way. Instructions such as '... and now take a superposition of all numbers from the previous operations...' are meaningless for any classical data processing device, but they make a great deal of sense to a quantum computer, and give it revolutionary processing power.

As a result of quantum computation we can construct new ways of solving problems, some of which can turn formidable mathematical procedures - such as factorization - into easy ones.

The difficulty of factorization underpins the security of many common methods of encryption; for example, RSA (the most popular public key cryptosystem) gets its security from the difficulty of factoring large numbers.

In 1994, Peter Shor from AT&T's Bell Laboratories in New Jersey devised the first 'killer application' of quantum computers: a quantum algorithm that allows a quantum computer to perform efficient factorization. For over a decade the whole issue of quantum computation had seemed little more than an academic curiosity. Shor's algorithm - which illustrated the potential of quantum computation - has stimulated research efforts world-wide.

3.2 Algorithms and the Complexity Problem

In order to solve a particular problem computers follow a precise set of instructions that can be mechanically applied to yield the solution to any given instance of the problem. A specification of this set of instructions is called an algorithm. Examples of algorithms are the procedures taught in elementary schools for adding and multiplying whole numbers; when these procedures are mechanically applied, they always yield the correct result for any pair of whole numbers. Some algorithms are fast (e.g. multiplication) others are very slow (e.g. factorization, playing chess). Consider, for example, the following factorization problem:

? x ? = 29083

How long would it take, using paper and pencil, to find the two whole numbers which should be written instead of the two question marks (the solution is unique)? Probably about one hour. Solving the reverse problem:

127 x 229 = ?

again using paper and pencil technique, takes less than a minute. All because we know fast algorithms for multiplication but we do not know equally fast ones for factorization.

What really counts for a "fast" or a "usable" algorithm, according to the standard definition, is not the actual time taken to multiply a particular pair of numbers but the fact that the time does not increase too sharply when we apply the same method to ever larger numbers.

Skipping details of the computational complexity, computer scientists have a rigorous way of defining what makes an algorithm fast (and usable) or slow (and unusable). For an algorithm to be fast, the time it takes to execute the algorithm must increase no faster than a polynomial function of the size of the input.

The input size can be thought of as the total number of bits needed to specify the input to the problem - for example, the number of bits needed to encode the number we want to factorize.

If the best algorithm we know for a particular problem has the execution time (viewed as a function of the size of the input) bounded by a polynomial then we say that the problem belongs to class P. Problems outside class P are known as hard problems.

Thus we say, for example, that multiplication is in P whereas factorization is not in P and that is why it is a hard problem. Hard does not mean "impossible to solve" or "non-computable" - factorization is perfectly computable using a classical computer. However, the physical resources needed to factor a large number are such that, for all practical purposes, it can be regarded as intractable on classical machines.

It is worth pointing out that computer scientists have carefully constructed the definitions of efficient and inefficient algorithms trying to avoid any reference to a physical hardware. According to the above definition factorization is a hard problem for any classical computer regardless of its make and clock-speed.

The same standard text-book method of multiplication requires little extra work when we switch from two three digit numbers to two thirty digits numbers. By contrast, factoring a 30 digit number using the simplest trial division method is about 1000 times more time (or memory) consuming than factoring a three digit number. The use of computational resources is enormous when we keep increasing the number of digits.

The largest number that has been factorized as a mathematical challenge, i.e. a number whose factors were secretly chosen by mathematicians in order to present a challenge to other mathematicians, had 129 digits. No one can even conceive of how one might factorize, say, thousand-digit numbers using a classical computer; the computation would take much more that the estimated age of the universe.

3.3 The Quantum Computation Answer

Consider a register (used to hold data in computer memories) composed of three physical bits. Any classical register of that type can store in a given moment of time only one out of eight different numbers i.e. the register can be in only one out of eight possible configurations such as 000, 001, 010, ... 111.

But a quantum register composed of three qubits can store, in a given moment of time, all eight numbers in a quantum superposition.

If we keep adding qubits to the register we increase its storage capacity exponentially i.e. three qubits can store 8 different numbers at once, four qubits can store 16 different numbers at once, and so on; in general L qubits can store 2^L numbers at once.

Once the register is prepared in a superposition of different numbers we can perform simultaneous operations on all of them. For example, if qubits are atoms then suitably tuned laser pulses affect atomic electronic states and evolve initial superpositions of encoded numbers into different superpositions. During such evolution each number in the superposition is affected and as the result we generate a massive parallel computation contained within one piece of quantum hardware.

This means that a quantum computer can - in only one computational step - perform the same mathematical operation on 2^L different input numbers encoded in coherent superpositions of L qubits. In order to accomplish the same task any classical computer has to repeat the same computation 2^L times or one has to use 2^L different processors working in parallel. In other words a quantum computer offers an enormous gain in the use of computational resources such as time and memory.

Although this sounds like yet another purely technological progress (with classical computers doing the same computations as quantum computers, but simply needing more time or more memory), the important distinction to make is that classical computers need *exponentially* more time or memory to match the power of quantum computers.

Purely technological progress will never provide exponential speed-up; it can only increase the computational speed by a fixed multiplicative factor. That does not help to change the exponential dependence between the size of the input and the execution time.

Such change requires inventing new, better algorithms. Although quantum computation requires new quantum technology, its real power lies in new quantum algorithms which exploit quantum superposition. This is the phenomenon that can contain an exponential number of different terms and thus beat the exponential dependence.

3.4 Quantum Algorithms

Shor's algorithm, mentioned in the introduction to this section, gives only one example of the tasks which a quantum computer might accomplish. Although the story of quantum computation started as early as 1982, when the physicist Richard Feynman considered simulation of quantum-mechanical objects by other quantum systems, the unusual power of quantum computation was not really anticipated until 1985. It was then that David Deutsch of the University of Oxford published a crucial theoretical paper in which he described a universal quantum computer and what it might achieve.

Deutsch indicated that quantum computers could perform certain computational tasks more efficiently than classical computers. He proposed a scheme which calculates the value of a function for two different inputs, and allows the two values to be compared. This comparison is achieved using only a single function evaluation which is applied to both inputs simultaneously.

Although it is of great historical interest, this idea has few practical implications. After the Deutsch paper, the hunt was on for something interesting for quantum computers to do. At the time all that could be found were a few rather contrived mathematical problems.

This all changed rather suddenly in 1994 when Peter Shor devised his algorithm designed to perform efficient factorization, stimulating renewed worldwide interest in the possibilities of quantum computing. Generally, however, research has not produced an enormous number of useful algorithms.

A major step forward took place in 1997 when Lov Grover of AT&T Bell Laboratories published his remarkable searching algorithm. Using quantum interference effects to push the required information to the output, Grover's Algorithm is able to search a database containing N entries in approximately $\sqrt{N}$ steps.

To extract a specific telephone number from a database of a million entries, for example, the computer is able to examine all entries in the database at once, and then employ 1000 quantum interference operations to spread the specific number information into every branch of the entangled system. A measurement of the computer's quantum state will then reveal the required data.

A classical computer would have to scan the entries one by one and compare them to the required information. This would require, on average, 50,000 accesses to the computer's memory.

Variations on Grover's algorithm can perform a number of different search types. It could also make a formidable chess opponent, able to perform analysis of move sequences a billion times faster than a classical machine. This type of application will also make Grover's algorithm an important tool in cryptoanalysis. Classical cryptographic schemes such as DES (the Data Encryption Standard) can be cracked by searching among 10^{16} possible keys. Checking at a rate of one million keys per second, a classical computer would take over a thousand years. A quantum computer employing Grover's algorithm could radically improve this situation.

As an illustration of the power and relevance of quantum computing, quantum algorithms are undoubtedly impressive. There can be no doubt that more algorithms will be discovered as research continues, and it is likely that their usefulness will parallel that of Shor's and Grover's contributions.

Useful implementation of these algorithms relies, of course, on the implementation of quantum logic gates and networks, which will be discussed in the next section. However, it is worth pointing out that Feynman's idea of quantum simulations has been implemented in simple cases, Deutsch's problem has been implemented on NMR computers and Grover's algorithm is likely to be implemented in the foreseeable future, albeit with 'toy' data comprising three qubits.

3.5 Quantum Logic Gates and Networks

In principle, we know how to build a quantum computer; we start with simple quantum logic gates and connect them up into quantum networks. A quantum logic gate, like a classical gate, is a very simple computing device that performs one elementary quantum operation, usually on two qubits, (which can each be labeled as '0' or '1') in a given time.

A general quantum computation requires operations that involve one, two or more qubits at a time. In the course of such operations, qubits change their logical values in ways that depend on the states of other qubits in the register. It turns out that single-qubit gates, together with very simple two-qubit gates known as the quantum controlled-NOTs, suffice to implement any quantum network for any quantum computation.

NOT gates

Label the ground state of an ion, for example (see Chapter 4), as logical 0 and any other state of definite energy as logical 1. Suppose that the ion is initially in state 0 and that we want to effect the NOT operation, which just negates the logical value: 0 becomes 1, and 1 becomes 0. This can be done by shining a pulse of light of appropriate intensity and duration, with frequency corresponding to the energy difference between the two energy levels. This will cause the electron initially in the ground state 0 to evolve into state 1. If the electron were initially in state 1, the same light pulse would cause it to evolve to state 0. This logical NOT operation, though performed here by quantum means, can be understood in completely classical terms.

Of course, quantum logic gates differ from their classical counterparts in that they can create, and perform, operations on quantum superpositions; these operations have no classical analogues.

For instance, by using a light pulse with half the duration of the one needed to perform the NOT operation, we effect a half-flip between the two logical states. That means that the state of the ion after the pulse is neither 0 nor 1 but a coherent quantum superposition of the two.

Acting on the ion in that state with a further pulse of the same type will complete the NOT operation, and evolve the ion to state 1. Because the NOT operation can be performed as the composition of two such quantum "half-flips", this purely quantum operation is called the 'square root of NOT'. It is used in many quantum algorithms to create superpositions.

Controlled-NOT gate

The controlled-NOT gate acts on two qubits at a time, one of them known as the control qubit and the other as the target qubit. The logical value of the target qubit is negated if the control qubit is in state 1 and is left unchanged if the control qubit is in state 0. The logical value of the control qubit is always left unchanged. When the control qubit starts in a coherent superposition of logical values 0 and 1, the whole system of two qubits ends up in the corresponding superposition: the control qubit in state 0 with the target qubit in the initial state, and the control qubit in state 1 with the target qubit in the negation of the initial state. This type of state is known as an 'entangled state', and the two qubits are said to be 'entangled', because the pair as a whole contains more quantum information than is contained in maximal descriptions of each of them separately. Qubits typically become highly entangled at intermediate stages of a quantum computation. Entangled pairs of qubits are a valuable physical resource in many applications of quantum information theory.

Good quantum technologies are all those technologies which allow one to control quantum entanglement of two or more qubits (the more the better). This requirement boils down to implementing the quantum controlled-NOT gate.

3.6 Obstacles

As the number of quantum gates in a network increases, we quickly run into some serious practical problems. The more interacting qubits are involved, the harder it tends to be to engineer the interaction that would display the quantum interference necessary for data processing and extraction.

Apart from the technical difficulties of working at single-atom and single-photon scales, one of the most important problems is that of preventing the surrounding environment from being affected by the interactions that generate quantum superpositions.

The more components there are, the more likely it is that quantum information will spread outside the quantum computer and be lost into the environment, thus spoiling the computation. This process is called decoherence. Thus our task is to engineer sub-microscopic systems in which qubits interact with each other but not with the environment.

3.7 A Workable Solution: Quantum Error Correction

When we analyze physically realizable computations we have to consider errors which are due to decoherence - the computer-environment coupling. Considering the computational complexity which will be necessary to make quantum computing truly revolutionary, we need to assess how these errors scale with the input size.

If the probability of an error in a single run grows exponentially with the input size, then the randomized algorithm cannot technically be regarded as efficient any more, regardless of how weak the coupling to the environment may be.

Unfortunately the computer-environment interaction leads to just such an unwelcome exponential increase of the error rate with the input size. It is clear that for quantum computation of any reasonable length to ever be physically feasible, it will be necessary to incorporate some efficiently realizable stabilization scheme to combat the effects of decoherence.

In 1993 Deutsch discussed this problem (during the Rank Prize Funds Mini-Symposium on Quantum Communication and Cryptography, Broadway, England) and proposed 'recoherence' based on a symmetrisation procedure.

The basic idea is to control not one but several quantum systems (computers) which are initially prepared in the same quantum state. If there are no errors then each quantum system evolves independently, and at each moment of time all systems are in the same state. The total system (composed of all the systems in question) is in a very symmetric state. If one of the systems evolves differently, due to errors, then the symmetry breaks down; this can be corrected by examination of the other inerrant systems.

Quantum error correction was given a more formal framework in 1995 when Peter Shor and Andrew Steane of Oxford University independently proposed the idea of error correcting codes. In this system a single qubit is encoded by a string of qubits. The encoding is carried out such that an error in a single qubit will still allow the original qubit to be inferred from the code string and thus restored.

All of the operations required for this process can be arranged so that they introduce less noise than they suppress. The idea of introducing extra quantum information processing in order to compensate for the imperfections inherent in such operations may seem counterintuitive, but it has been proven to be a worthwhile approach.

A quantum computer performing a task beyond the capabilities of classical computers (say 1000 qubits and more than a billion quantum logic gates) would require environmental noise levels so low that they are considered unattainable. But the introduction of error correction routines would render such a computation possible. Although it might require a hundred-fold increase in the size of the computer, and thousands of extra gates for error correction at each stage of the computation, the noise levels at which such a machine could operate are not beyond our reach.

Although the field of quantum error correction is extremely new, its worth has already been well proven. The attention it has received from many of the most ingenious minds of the quantum information community has ensured that the field has made significant advances since it was first discussed.

Quantum error correction will allow qubits involved in a long quantum computation to be 'lost' as many as a million times during the lifetime of the operation. Through the application of this powerful notion we can confidently say that decoherence, the main obstacle to quantum computation, is not an invincible adversary.

3.8 Conclusions

We have demonstrated the power and the principles that lie behind quantum computing. Everything we have discussed is (in principle) possible. The algorithms, and the logic gates necessary for their implementation, offer a means to achieve the exponential speedup that is necessary to move beyond the limitations of classical computing.

The current challenge is not to build a fully-fledged universal quantum computer right away, but rather to move from the experiments in which we merely observe quantum phenomena to experiments in which we can control those phenomena in the necessary ways. Any quantum system with at least two discrete quantum states (e.g. energy levels) could be used as a qubit. The most promising candidates so far have been polarized photons (for instance, in quantum cryptography), nuclear spins (in the NMR quantum computation), and trapped ions. Simple quantum logic gates involving two qubits are now being realized in laboratories in Europe and the U.S.A.

The emerging concept of quantum error correction has shown that the main obstacles to the implementation of quantum computing are not insurmountable. It would seem, therefore, that there are no fundamental reasons why quantum computing is not possible. The questions that remain have far less to do with 'if' than 'when' it will happen.

Chapter 4
Experimental Realizations

Michael Brooks

4.1 Introduction

Currently, research is being undertaken on a number of possibilities for the implementation of the logic gates and networks discussed in the previous section, with the ultimate goal of implementing the remarkable algorithms that have been mentioned. However, all the technologies mentioned below will also have implications for quantum information processing in general. Further research may uncover other technologies which may provide an easier route for the implementation of quantum information processing.

4.2 Trapped Ions

A collection of ions stored in a linear trap provides one of the more promising schemes for building a quantum computer, with each qubit (one per ion) being comprised of a pair of the ions' internal states. All ions in the trap have the same charge and repel each other, so any motion of one of the ions is transferred by this electrostatic repulsion to other ions in the trap, inducing various collective motions known as phonons. A single ion can be set in motion by directing a laser pulse at that particular ion; because the inter-ion separation is much larger then the wavelength that induces the excitation, no other ions are affected, allowing the realization of one-bit gates. Ignacio Cirac and Peter Zoller of the University of Innsbruck first proposed this system, which has the great advantage of being scalable, in principle, to an arbitrary number of qubits.

The Coulomb repulsion between the charged ions provides a physical mechanism to implement a Controlled-NOT gate between a pair of qubits. Quantum information is transferred between different ions using a particular quantized mode of the phonon (the ions' collective motion) - this plays the role of a 'quantum data bus'. The presence or absence of the phonon affects the energy levels of the ions, allowing the target ion to respond differently to light according to the state of the control ion. Using this idea, the qubit stored in one of the ions can be transferred to the collective phonon mode by directing a laser pulse at that particular ion. The presence or absence of such a phonon affects the occupation of energy levels of the other ions, allowing the target ion to respond differently to light according to the state of the control ion. A sequence of very accurately tuned pulses aimed at the control and target ions allows the two ions to interact only with each other, so that we can arrange for the control to remain unchanged and for the state of the target to be negated only when the control is in state 1.

Before the quantum gate can be implemented, the phonon mode must be initialized in a pure quantum state - this requires laser cooling of the ions to their motional ground state. Finally, a high efficiency read-out of the qubit is implemented using the method of quantum jumps.

These theoretical ideas are currently being implemented by experimental groups both in Europe (Innsbruck, Munich, Aarhus and Oxford) and in the U.S.A (NIST Boulder, Los Alamos); the basic techniques of controlled storage and laser cooling of single and several ions in traps have been, and are presently developed as, a standard tool in these laboratories. The basics of the Controlled-NOT operation of the ion trap quantum computer scheme have been demonstrated first by the NIST Boulder group, operating between the motional mode of a single trapped Beryllium ion and two of the ion's internal hyperfine structure states. In this experiment the motional state was initialized in the ground state by side band laser. Ground cooling of the collective modes of two Be ions, which is a prerequisite of demonstrating a universal two-bit quantum gate, has been achieved by the NIST Boulder group. Recently, Q.A. Turchette et al. have reported the generation of Bell-like singlet and triplet entangled states. In contrast to previous experiments with entangled states with either massive particles or photons this entanglement was produced in a deterministic fashion, producing entanglement on demand without selection, thus demonstrating a crucial building block for large scale quantum computation.

In present experiments, decoherence times of the qubits stored in the internal ground states of the ions have been demonstrated to be of the order of thousands of seconds, limited only by collisions with background atoms. The dominant decoherence mechanism at present is heating of the phonon modes, whose physical mechanism is not currently understood in detail. However, recent experiments with two ions are encouraging: the heating of the modes of relative ion motion is substantially suppressed relative to the center-of-mass mode (with decoherence rates less than 0.1 ms^{-1}), suggesting the importance of these modes in future experiments. Addressing of the individual ions, in particular in tight traps with small separation between the ions, remains a technical challenge, although exploiting the micromotion of the ions has allowed the control of the Rabi frequency at least for two ions in the recent Boulder experiments.

Starting from the present experimental situation, one expects that small model systems of ion trap quantum computers with up to a few tens of ions can be built in the next ten years, and it is worth re-stating that an ion trap quantum computer has no fundamental scaling limits, though, in order to realise ion trap quantum computers beyond about 10 qubits, fundamental breakthroughs in the control of the decoherence mechanism are essential. Future generations of experiments will also involve the coupling of ion traps using cavity QED, either as a way to implement distributed quantum computing, or in the context of quantum communications over long distances.

4.3 Nuclear Magnetic Resonance

In Nuclear Magnetic Resonance computers the two spin states of an atomic nucleus in a magnetic field are used as the two states of a qubit. Different atoms in a molecule can be distinguished, and so a molecule can be used as a quantum computer, with each nucleus providing a single qubit. Simple logic gates which only affect a single qubit are easily implemented using radio frequency fields. These interact strongly with nuclear spins, allowing them to be controlled with great precision. In order to perform interesting computations, however, more complex gates are needed, which allow the state of one qubit to affect other qubits in the computer. This requires some form of interaction between nuclear spins, so that one spin can sense the state of other spins in the molecule. This is easily achieved, as the naturally occurring spin-spin coupling interaction has the desired form.

NMR computers are different from other quantum computers in one important respect. The NMR signal from a single molecule is far too weak to be detected and so it is necessary to use a large number of identical copies to amplify the signal. This is not difficult as even a few milligrams of a chemical compound will contain the required number of molecules. It is, however, impossible to ensure that all the copies start the calculation in the same initial state, and so different copies will in effect perform different calculations, making it extremely difficult to extract the desired result. This inability to prepare NMR computers in a well defined initial state prevented their use for many years.

In 1997, two separate solutions to this problem were published. Both papers described how to 'distil' an effectively pure starting state from a complex mixture, and thus run the calculation of interest (the signals from other starting states can be arranged to cancel, so that they make no overall contribution to the final result). Since then progress has been extremely rapid. Two different two-qubit NMR computers have been built: one by Chuang and coworkers, based on the ^{1}H and ^{13}C nuclei in isotopically labelled chloroform, and one at Oxford University using two ^{1}H nuclei in cytosine. Both systems have been used to implement a simple quantum algorithm to solve Deutsch's problem, which calculates the value of a function for two different inputs, and allows the two values to be compared. This comparison is achieved using only a single function evaluation which is applied to both inputs simultaneously.

Deutsch's problem is of great historical interest, as it shows that quantum computers can perform calculations which are literally impossible for any classical computer, but it has few practical implications. By contrast, Grover's quantum search algorithms have great practical significance. The simplest version, searching for a single item among four possibilities, can be implemented on a two qubit computer, and is thus within the range of current NMR systems. Chuang and coworkers have recently described a complete implementation of this algorithm on their chloroform computer. Their results show that it is possible to locate the desired item in one step, as predicted, and that it is also possible to determine that there is indeed only one item to be found. A similar but less complete implementation of this algorithm has been performed on the cytosine computer at Oxford University.

Implementing a quantum search algorithm is a major step forward for NMR quantum computers, but by no means the limit of their development. The next step is to implement more complex algorithms on larger systems (preliminary results on three qubit systems have already been demonstrated). Some commentators have claimed that it will not be possible to build NMR computers with large numbers of qubits, as the efficiency of the distillation process used to obtain an initial pure state falls off rapidly with the number of qubits. This criticism is almost certainly correct, but in practice is unlikely to prove a significant limitation in the near future. Other problems, related to the difficulty of selectively addressing individual spins, will make it difficult to build NMR computers with more than about six qubits. This would, however, provide the ability to investigate many important problems. It is also much larger than any quantum computer likely to be built using other approaches in the near future. NMR is likely to remain a leading technology for quantum computers for many years to come.

4.4 Cavity Quantum Electrodynamics

Cavity quantum electrodynamics is a possibility for qubit coding and processing, although it is experimentally extremely challenging. Several experimental set-ups that were initially designed to test fundamental aspects of quantum theory have been used recently to demonstrate quantum logic operations, or quantum memories. The basic premise is that atoms are held in an optical cavity (a region of space bounded by reflecting surfaces). If the wavelength of light is similar to the dimensions of the cavity, the light can be used to couple to and control the quantum states of the atom, providing a way to encode or process information. The atom provides an index of refraction for the field that is large enough to produce macroscopic effects. The photons can transfer their entanglement onto atoms in the cavity, which can then be used to store the quantum information while the photons can be used to transmit it between atoms or ions. A group in the USA (Caltech) is close to demonstrating conditional logic at the single quantum level, and several European laboratories have carried out seminal work in this area. The Paris group has entangled very different quantum states of atoms and fields, as well as atom-atom entanglements. They have also demonstrated for the first time the effect of controlled decoherence on a quantum state, and are currently working on field-field entanglements.

4.5 Quantum Dots

Solid state realisations of quantum gates and quantum computation principles are presently studied by a number of groups. One possibility is to use the technologies developed for quantum dots. These are regions of semiconductor on the nanometer scale, contained within bulk semiconductors, which confine the motion of electrons very narrowly such that the electrons can only exist in a few well-defined quantum states. Working with semiconductors is now a mature technology, and this approach offers the prospect of a new source of single-photon pulses for quantum communications. It also holds some potential for encoding and storing quantum information. In the long run, the possibilities of implementing quantum logic gates by having neighbouring quantum dots interact should certainly be investigated. Another possible solid state implementation includes quantum computation using superconductors and Josephson junctions. A group at Jyvaeskylae in Finland is heavily engaged in studying these possibilities for the implementation of quantum logic gates. Finally, the realisation of various single-electron phenomena in micro-electronic structures, e.g., the single-electron turnstile, should certainly be investigated with respect to its possibilities for quantum computation.

Chapter 5
Optical Technologies for Quantum Computing and Communications

Michael Brooks

5.1 Introduction

In discussing the technology that can be developed for QCC, a useful starting point is to examine the 'quantum engineering' solutions that have been proposed in the field of optical telecommunications. In this field, the information is not encoded in terms of qubits that can take both values of 0 and 1 simultaneously, but rather as 'classical' bits, each having a definite value of 0 or 1. The aim of quantum engineering then is to preserve the classical value assigned to the bit in spite of the uncertainty that quantum mechanics inevitably introduces at every stage of the life of the optical signal: its generation, its transmission, its amplification, its distribution or its detection.

The engineering of the quantum uncertainty of classical bits in optical communications (whether it be over thousands of kilometers, such as in transoceanic fiber communications, or over a few microns, as in intra-chip optical signal processing) can be considered as the 'first generation' of quantum information technology. It is presently a very active field, that emerged only 10 or 12 years ago and in which several European laboratories have pioneered and are still producing innovative contributions. This 'first generation' technology exploits the quantum mechanical correlations that specific nonlinear optical processes impart on photons to manipulate the quantum mechanical uncertainty.

In a 'second generation' of quantum information technology, these subtle quantum mechanical operations can be extended to the manipulation of the quantum mechanical superpositions that constitute the qubits. Then, the information itself will be coded in a quantum state, in view of developing quantum computation.

5.2 Using the Quantum Nature of Light

5.2.1 Potential

Light is a familiar 'physical object' that can be considered as the prototype of a quantum system for coding information. As a wave, light possesses a large number of observables into which information can be coded, such as amplitude and phase, quadratures (i.e. in-phase and 90 degrees out-of-phase with respect to a reference oscillator), frequency and duration, polarization, etc.

In optical telecommunications, binary (or analogue) information is coded in the intensity, the phase or the polarization of a pulse of light. This pulse is generally produced by a laser, is then transmitted through transparent medium (usually an optical fiber) and is subsequently detected by a photodiode that produces an electrical signal (proportional to the energy content of the pulse) for further processing.

One important feature that needs to be taken into consideration when coding information in a quantum system is that, according to quantum mechanics, the values of all observables of the system may fluctuate randomly. However, there are pairs of observables for which the quantum fluctuations of the two members are related: when one decreases, the other increases. Thus, it is possible to fix (or measure) the value of one observable to the desired accuracy at the expense of accepting increased fluctuations for its conjugate observable. This feature is at the root of all strategies of quantum processing of information as it dictates the way that we must read, write, store, duplicate or transmit information in a way that permits us to have the laws of quantum mechanics on our side.

5.2.2 Problems

While in classical physics there is nothing wrong with a photodiode measuring the energy of a laser pulse, quantum mechanics tells us that things are not as simple as they look. An ideal laser emits 'coherent' light, that is a wave that has a relatively well-defined amplitude and phase, while a photodiode detects energy, that is the number of photons that is incident on it. Thus, the process of pulse generation and the process of pulse detection deal with two different variables (amplitude and photon number) which, according to quantum mechanics, are not compatible. If we want to express the coherent light of the laser in terms of photon numbers, which the detector can 'see', a laser pulse will consist of not only the photon number state that corresponds to its nominal energy content, but will actually be a superposition of all possible photon number states.

Let us examine briefly in terms of a concrete example what are the implications of this quantum mechanical superposition. As an example, a pulse of coherent light that contains nominally 10 photons corresponds to a superposition of all photon number states in which the 10-photon state has a weight of only 12.5%. The histogram of probabilities of finding a given photon number in this superposition extends from the zero-photon state (with a weight of 0.0045%) all the way to infinity; it peaks at 10 photons and has a width (standard deviation) of the order of plus or minus 3.16 photons (i.e. the square root of 10). This means that the nominal number of photons that compose the pulse (or, equivalently, its energy content) has a meaning only in the sense of a 'most probable' value. An actual measurement may give any value within the superposition, including the value of zero photons. This uncertainty in the result of the energy measurement of a laser pulse that is due to the incompatibility between the coherent amplitude and the photon number is often considered by telecommunication engineers as a 'quantum noise' that contaminates the nominal value of the signal.

5.3 Quantum Noise in Optical Communications

The quantum noise distribution whereby the width corresponds to the square root of the mean is often called the 'Standard Quantum Limit' (SQL) and represents the ultimate noise level in a pulse of coherent light, once all the technical sources of noise have been overcome. One useful quantity used by the engineers is the Signal-to-Noise Ratio (SNR) which can be taken as a measure of the ease with which we can read a signal in the presence of noise. At the SQL the SNR is proportional to the mean photon number.

The existence of the SQL introduces the possibility of an error when reading the bit value that is coded by a given pulse. For example, if we use 10-photon pulses to code the binary value 1 there is a finite probability that an energy measurement of a 10-photon pulse may give zero (or one or two) photons, and thus produce an error in which a 1-bit is mistaken for a 0-bit. Clearly, because the SNR goes as the square root of the mean value, the more intense is the pulse used for coding the value 1, the larger the SNR, thus resulting in a smaller Bit-Error-Rate (BER). Nevertheless, even if at the laser source we start out with a pulse that is quite intense (and thus has a small relative quantum noise) the subsequent operations that are performed on the pulse will inevitably increase its quantum noise and thus degrade the SNR and the BER. In what follows, we shall briefly examine some of the operations that a signal pulse undergoes in the course of its life in an optical communication system, the incidence of these operations on the quantum noise, and the quantum engineering solutions that have been implemented in order to circumvent these problems. Firstly, though, we will take a look at the necessary technologies.

5.4 Generic Technologies in Quantum Communications

As we have seen, in present-day optical communications information is coded classically onto the optical signal. Nevertheless, the quantum mechanical nature of the information coding process makes itself felt through the quantum fluctuations to which the signal is subject. These quantum fluctuations can be tackled through the development of specific quantum optical devices based on two generic technologies, Nonlinear Optics and Cavity Quantum Electrodynamics, which permit one to introduce quantum correlations among the photons thus giving rise to quantum processing of 'classical information'.

5.4.1 Nonlinear Optics

In our everyday experience, light does not interact with light, and photons can traverse each other without changing their characteristics, such as their color or direction. We live in a world of 'linear optics'. There are special materials, however, inside which laser pulses can interact with each other changing their frequency, their speed or their direction. This field, in which light can interact with light on a special material support, is called nonlinear optics.

The interaction of photons with each other imparts quantum mechanical correlations that can be exploited to achieve quantum mechanical processing of optical signals. Among the very large variety of nonlinear optical phenomena, two have been implemented more particularly in quantum information processing.

The first one, parametric downconversion, is a three-wave mixing process. In this process, an incident photon undergoes fission to produce two twin photons of half the energy. The quantum fluctuations of the two emerging modes of the electromagnetic field have quantum mechanical correlations which are exploited in achieving noiseless amplification or noiseless copying of optical information.

The second nonlinear optical process is nonlinear refraction, which is a four-wave mixing process. In nonlinear refraction, a propagating laser beam modifies momentarily the refractive index of the medium, and thus can change its own speed of propagation (self-phase modulation) or the speed of another co-propagating beam (cross-phase modulation).

5.4.2 Cavity Quantum Electrodynamics

Optical cavities are regions of space bounded by reflecting surfaces, so that light can be confined inside them. When the characteristic dimensions of the confinement are of the order of the wavelength of light, then the interaction of the radiation at the corresponding frequencies with any material system present in the cavity is modified, and novel manifestations of the light-matter interaction may arise. In particular, quantum mechanical manifestations, such as the spontaneous emission of light, can be inhibited or enhanced, while a transparent material placed in the cavity could display optical nonlinearities (such as frequency doubling or a nonlinear refractive index) that may be manifested with intensities corresponding to only a few photons. Cavity quantum electrodynamic phenomena are presently being exploited to reduce the quantum fluctuations of light sources useable in optical communications.

The area of cavity quantum electrodynamics has been active for the past fifteen years or so and several European laboratories have produced seminal contributions in this field. The more specific line of research in which cavity quantum electrodynamics is being applied to semiconductor optoelectronics was launched about five years ago, with several European groups leading the way and striving towards the development of a functioning optoelectronic device.

5.5 Operations Performed on Optical Signals

5.5.1 Signal Generation

The light signals that are used in optical communications come from semiconductor lasers, whose operation is based on the principle of stimulated emission. In stimulated emission photons are emitted in a controlled way, all with the same characteristics, and thus give rise to a coherent beam. However, in addition to the stimulated coherent emission the active medium of the laser emits photons spontaneously, in all directions and in an uncontrolled way, because of the quantum nature of the radiative interaction. This uncontrolled spontaneous emission represents an energy loss mechanism as well as a source of excess quantum noise in the output signal.

Solution Strategy I: Microcavities and Cavity QED

In order to overcome the problems introduced by spontaneous emission in lasers, a very active area of quantum engineering has been opened up recently aiming at inhibiting spontaneous emission in the unwanted directions and reinforcing it in the useful output of the device. This is done by use of optical cavities whose characteristic dimensions are of the order of the wavelength of light (microcavities).
Microcavities operate according to the principles of cavity quantum electrodynamics whereby the interaction between light and matter as well as the quantum fluctuations of light can be modified or controlled by the appropriate engineering of the cavity geometry.

Solution Strategy II: Squeezed States

Another aspect of noise reduction that 10 years ago mobilized a large effort in quantum engineering, is the production of 'squeezed states' of light, by use of nonlinear optical interactions. In these quantum states of light the noise of the radiation field is reduced in one quadrature of oscillation below the standard quantum limit, at the expense of the other quadrature in which noise is increased. If the information is coded in the low-noise quadrature, then it can be read in a homodyne detection scheme with a SNR better than that permitted by the SQL, resulting in an improvement of the BER. Unfortunately, squeezed states have proved to be difficult to make and even more difficult to maintain, since they lose their low-noise properties when they undergo attenuation brought about by the inevitable in-line losses.

5.5.2 Detection

Quantum mechanics tell us that we cannot perform a measurement without perturbing the system we are measuring. An extreme case of that statement is the detection of a photon, which involves its conversion into an electron by means of the photoelectric effect. Clearly, in the measurement of the energy of a light pulse the 'perturbation' is its complete destruction, since the light pulse does not survive a measurement of its energy. We may try to minimize this perturbation by introducing a branching point in the path of the light pulse and thus sample only some of its energy. However, as we shall see below when we discuss branching, this procedure increases the fluctuations and the error rate of both branches. In addition, in the part of the pulse that is not killed by the measurement, there is an increase of the relative quantum noise and thus a subsequent measurement will necessarily produce a higher error rate. There are, of course, many instances in which the same information must be read repeatedly, each time with the same accuracy, and in this case sampling is no longer an acceptable solution. To be able to read optical information repeatedly, we have to be able to 'see a photon without killing it'. This challenge is met by specific quantum mechanical strategies known under the term 'Quantum Non-Demolition' measurements.

Solution Strategy I: Quantum Non-Demolition

In QND measurements, the information that is coded in a light beam is copied via a non-linear optical process onto a second light beam. This second beam can then be measured destructively, while the first beam can survive the measurement process and continue on its way unharmed. However, there is an additional subtlety here, because the copying process itself involves a quantum noise penalty. If this quantum noise is uncontrolled it can contaminate the original beam and thus the measurement can never be repeated with the same accuracy. The QND strategy consists of maintaining the information-carrying observable of the first beam noise-free and sending all the noise onto the conjugate observable which carries no information. In this way the laws of quantum mechanics are satisfied while at the same time the information-carrying channel is protected.

5.5.3 Attenuation

When a laser pulse propagates in an optical fiber, it loses gradually its energy due to the residual absorption in the fiber. This produces an attenuation of the pulse by a factor of 10 for every 50 km of propagation, and thus reduces progressively the mean value of photons in the pulse. As a consequence this reduces the quantum SNR by a factor of 3.16 every 50 km, and seriously degrades the BER. While the energy lost by attenuation can be re-injected in the signal by using special optical fibers that can amplify the signal, the degradation of the SNR is irreversible and cannot be overcome by further processing.

5.5.4 Distribution

Most optical transmission systems do not consist of a simple line linking a single emitter with a single detector, but are rather complex networks that include nodes and branching points in which the signal is divided into two or more channels. These channels may be transmission lines themselves or may be input/output ports into (and out of) the network. Upon branching, the energy is divided among the output channels and thus the SNR of the emerging pulses is degraded irreversibly with respect to that of the incoming pulses. A serious consequence of this is that quantum noise accumulates as branching points are cascaded, and its level may soon become prohibitive for data transmission. Thus, because of quantum noise, a network can only have a limited number of branching points or read-out ports.

In order to overcome the branching bottleneck, two solutions have been implemented. The first solution relies on the nonlinear optical process of 'parametric emission', whereby a photon that is incident on a nonlinear crystal undergoes 'fission' to give rise to two photons of half the energy. When a parametric emitter is used as an amplifier, it has two output channels. These channels can be configured in such a way that one is used to amplify an incoming information-carrying pulse (and as we shall see below this amplification can be noiseless in the sense that it can preserve the SNR), while the other channel gives a duplicate signal for free, and thus noiseless branching can be achieved. The second solution, which is also implemented by use of a nonlinear optical process, relies on the technique of Quantum-Non-Demolition measurements. Branching is performed by copying the information of the first channel into the second channel, while at the same time the noise that results in the copying process is prevented from contaminating the observable that carries the information, but is sent to the conjugate observable. In both of these solutions branching points can be cascaded without degrading the SNR of the information-carrying observable, and thus relatively complicated networks can in principle be constructed.

5.5.5 Amplification

When a light pulse is too weak to be detected, energy can be injected into it by making it pass through an optical amplifier. This increase of the pulse energy, however, is also accompanied by an increase in its noise: amplification makes the quantum noise problem worse. If the input pulse has the minimum amount of quantum noise by being at the SQL, the output pulse will have a very large amount of excess quantum noise, well beyond the SQL. To give a concrete example, let us examine the amplification of our 10-photon pulse, which at the

SQL has a SNR of 3.16. Upon amplification by a factor of 10, we get a pulse containing nominally 100 photons. As the input quantum noise is also amplified, we cannot improve the SNR and thus at best we should conserve it. In fact, a laser-based amplifier actually degrades the SNR by a factor of 2, and thus, at the output, our 100-photon pulse will have a SNR of 1.58. Clearly, this SNR is six times worse than the SQL for a 100-photon pulse, which would correspond to a SNR of 10.

The excess noise in optical amplification comes from the existence of uncontrolled channels in the amplifier in which spontaneous emission can occur in a random way. The individual spontaneous emission events deplete randomly the energy stored in the amplifier and thus cause fluctuations of the gain. Clearly, in order to get around the problem of quantum fluctuations in amplification we must use an amplifier that has no uncontrolled channel so that no noise can make its gain fluctuate. One amplifier that fulfills this requirement is the 'degenerate parametric amplifier' that amplifies a signal by stimulated two-photon parametric emission. Such an amplifier has only two channels, corresponding to the two signal quadratures: when one quadrature is amplified, the second quadrature is de-amplified. The amplifier is therefore sensitive to the signal phase. As far as quantum noise is concerned, this amplifier has a clear advantage: in the first quadrature, the signal and the noise are amplified by the same gain, and thus the SNR is preserved to its original value; at the same time, in the second quadrature both the signal and the noise are de-amplified, and thus the overall output of the amplifier does not contain any excess noise.

5.6 Conclusions: Towards the Second Generation

A word of caution is necessary when we speak of 'generations' of quantum information technology. The two 'generations' we are discussing represent an advance of one on the other in terms of their conceptual basis, since the quantum treatment is extended from classical to quantum bits. However, in terms of technical maturity, even the 'first generation' is still in the stage of exploration of the basic physical principles and the development of the basic physical tools and prototypes. It is not yet at the stage of an 'industrial' technology in which products and applications are available. The 'second generation' is still in the process of defining and setting up its conceptual basis and is adapting many of the 'first generation' tools to its specific purposes.

The two generic technologies (Nonlinear Optics and Cavity QED) for quantum processing of classical information that are being explored in optical communications have recently started to be extended for dealing with quantum coding in the form of qubits and for processing quantum information. This extension relies on a conceptually simple (but often quite difficult technically) implementation of these same techniques to the single-photon level.

The nonlinear process that has been more widely used in experiments on quantum information is parametric emission. This process produces two twin photons in an entangled state in which the quantum information is shared by the twins. When the twin photons are separated spatially, they give rise to non-local correlations. That is, the quantum information that is read in one of the two photons appears also 'instantaneously' in the other member of the pair, even when the two partners cannot communicate by any physical means (e.g. when they are separated by a distance that is larger than what could be bridged by the information travelling at the speed of light during the time it takes to perform the measurement). The non-local character of the twin-photon correlation was shown as early as 1982 for distances of the order of 15 meters, and more recently for a separation of the two photons of the order of 10 km. The entanglement of the twin photons and the non-locality of their correlations have found many applications in 'second-generation' quantum information experiments. One example is the setting up of a protocol for key transfer in quantum cryptography, where the two interlocutors share between them the two twins and establish a common random key by simultaneously reading the quantum information carried by the two photons. The use of twin photons is also under consideration in more complex protocols for the generation of cryptographic primitives based on bit commitment or on oblivious transfer. Another well-known example of putting the quantum entanglement of twin photons at work is the realization of quantum teleportation in which the quantum state of an object (a photon) is reconstructed on another object 'instantaneously', a large distance away.

One 'first generation' concept that will prove to be indispensable in developing 'second generation' quantum information processing systems is the concept of QND measurements, which are key elements for any quantum mechanical input-output functionality. It is clear, for example, that when a quantum memory has to be read repeatedly, this can only be done through a QND measurement process, which will not erase or alter the memory after the first reading. At the same time, the QND measurement paradigm, when performed on single photons, gives rise to controlled-not gates, in which the presence of a 'control' photon induces a change in the state of a 'target' photon. Such quantum mechanical gates would constitute key elements towards the construction of quantum-logic processing systems.

Regarding the second generic technology, cavity quantum electrodynamics, its relevant feature for developing 'second generation' quantum information processing systems is the strong interaction that exists between atoms and photons inside a cavity. This can be exploited in preparing quantum states of atoms and photons displaying entanglement in view of qubit coding and processing. Several experimental setups that were initially designed to test fundamental aspects of quantum theory have been used recently to demonstrate quantum logic operations, or quantum memories.

In addition, the two generic technologies can be combined in ways which take advantage of the strong points of each other. For example, CQED can be invoked in enhancing optical nonlinearities in different material systems in order to make these effects operational at the single-photon level. Also, the effects obtained from entangled twin photons coming from a parametric emitter can be combined with those obtained in CQED or in ion traps. Thus, the photons can transfer their entanglement onto atoms in a cavity or onto ions, the material particles can be used to store the quantum information while the photons can be used to transmit it between atoms or ions.

Finally, before ending our discussion on the prospects of a 'second generation' of quantum information processing systems, it is necessary to point out that CQED, ion traps or any other technology that relies on the quantum states of material particles, suffer from a limitation that is not shared by the nonlinear optical technologies. The excited states of material particles are in constant interaction with their environment and are thus extremely fragile. This interaction brings about a relatively rapid loss of the quantum information known as decoherence, which seriously limits the time during which quantum information can be processed effectively. Thus, although it may be possible to tolerate some decoherence when dealing with relatively small numbers of qubits (for example in the realization of cryptographic primitives) its presence becomes prohibitive whenever large numbers of qubits need to be processed and, at present, constitutes the main impediment towards the development of a 'second generation' of reliable quantum information processing systems.

Chapter 6
Applications

Michael Brooks

6.1 Introduction

It is worth remarking at the outset of this chapter that, in all likelihood, most of the applications of quantum information processing are not yet even dreamed of. When few-qubit quantum devices become technologically feasible - which is an absolute pre-requisite for the entire field of quantum computation - possibilities other than those considered here will undoubtedly appear.

The number of qubits whose coherence has to be maintained is the single most important factor in attempting to bring applications to the theory of quantum information processing. It determines the extent of the technological challenge posed by the application, and thus the probable time needed to develop appropriate solutions.

Even though it is still in its relative infancy, quantum information processing is already set to offer new technologies for use after the turn of the century. Applications which need only single qubits - such as quantum cryptography - could find their first niche markets in the very near future, possibly within 3-5 years. However, applications that require the simultaneous processing of thousands of qubits, such as quantum computers that are able to factorize large integers, are likely to take tens of years to reach operational maturity.

Quantum physics also has a great deal to offer the areas of metrology, sensing and simulation. Given that industrial research and development teams, and tomorrow's high-tech society, will require all of these fields to produce technologies that exhibit ever-better performance, it is clear that the motivation for the entire field of QIP is even broader and more promising than for a futuristic quantum computer.

6.2 Emerging Technologies

Some applications that could be developed in the medium-long term (5-20 years) are discussed below.

6.2.1 Quantum Cryptography

The Heisenberg uncertainty principle implies that whenever and however a quantity is measured with high, medium or low precision, 'noise is added' to the conjugate quantity, that is the conjugate quantity becomes more smeared. Hence, by encoding bits in conjugate variables, the basic laws of quantum physics guarantee that any measurement will cause a sizeable - thus measurable - 'added noise'. In particular if an unauthorised eavesdropper tries to extract information from such a quantum channel (i.e. a channel carrying a bit encoded in quantum states) she necessarily reduces the correlation between the input bits and the output bits. The legitimate users, known as Alice and Bob, can thus detect the presence of any eavesdropper. In order to detect the eavesdropper before sending information, the above idea can be complemented by a second idea: the quantum channel is not used to send messages (information), but only to send a key (random strings of bits without information). Hence, Alice and Bob can check whether the key is confidential (perfect correlation between their bits), or whether an eavesdropper's measurement caused "noise". All this can be done before they use the key to encode their message. In practice, Alice and Bob bits are never perfectly correlated, hence one has to find bounds on the possible amount of information that the eavesdropper could have for a given Quantum Bit Error Rate (QBER). The QBER can be measured by Alice and Bob. They need to first apply some error correction algorithm, next some privacy amplification scheme. At present, the experimental feasibility of quantum cryptography using standard telecom fibers over tens of km is well established, as demonstrated by BT, Los Alamos, Geneva University and more recently IBM; the dark counts of the photon counters is the major technical difficulty. Bit rates up to some Kb with QBER of a few percent should be achievable in industrial prototypes but the distances seem severely limited to about 100 km. The bounds on the eavesdropper information for given QBER is known only under some (very plausible) assumptions.

6.2.2 Quantum Repeaters

Quantum repeaters that could extend the distance of quantum cryptography links have been proposed, though no experiment has yet been realised (nor even seriously planned). The general idea is to exploit quantum nonlocality to establish a quantum link between two locations connected by a quantum communication channel in such a way that the quality of the link is higher than that of the channel. For example, the accumulation of noise in the channel typically imposes limits on the channel quality that explode exponentially with the distance. But quantum repeaters could establish links of arbitrary high quality, independently of the distance. In practice, such repeaters rely on purification schemes which allow users to prepare pairs of particles spatially separated, but with arbitrarily high quantum correlations.

The implication of realizations of quantum repeaters are huge and very serious. One could extend quantum cryptography to arbitrary distances. One could also extend other kinds of quantum communications, like teleportation, to arbitrary distances. Furthermore, one could put quantum computers in networks (e.g. many small quantum computers would form a 'real' - useful - quantum computer). It should be pointed out that these are still only theoretical ideas.

6.2.3 Quantum Simulators

Quantum computers are devices that process information stored on quantum systems such as atoms, photons, and nuclear spins. Unlike conventional computers, whose operation is inherently dissipative and destroys quantum coherence, quantum computers are able to process information in a way which preserves quantum phase information: this feature allows them to perform computations in ways that classical computers cannot. Perhaps the best-known example of a calculation that can be significantly speeded up by using a quantum computer is factoring: a large number of the form pq, where p and q are unknown primes can be rapidly factored using a quantum computer to reveal the values of p and q. Since the supposed difficulty of factoring underlies the security of current public key cryptosystems, whether or not quantum computers can be built is a matter of considerable interest. Other potential applications of quantum computers include database searching, and the determination of global properties of unknown functions.

The first proposed application of quantum computers was to a class of problems in physics: in 1982, Richard Feynman conjectured that quantum computers could be much more efficient than classical computers for the problem of simulating other quantum systems. Simulating quantum systems, e.g., by Monte Carlo techniques, is a difficult task that uses large amounts of computer time. The basic problem is that a quantum system with N degrees of freedom occupies a Hilbert space of size at least 2^N - even writing down the state of a quantum system requires an exponentially large amount of space on a classical computer - and evolving that state forward in time requires the exponentiation of 2^N by 2^N matrices. As a result, following the time evolution of a quantum system of, say, 40 spins is effectively impossible on today's computers and on computers for some time to come. In 1996, Lloyd showed that Feynman's conjecture was correct: quantum computers could function as universal quantum simulators, providing efficient analog simulations of any quantum system that evolved according to local interactions. These results were subsequently applied to fermionic systems, spin glasses, to molecular and atomic wavefunctions and spectra, and to quantum chaos: for such systems, a quantum computer with a few tens of spins could, in principle, provide more accurate simulations than current classical computers.

In 1998, David Cory provided an experimental demonstration of the power of quantum simulation: using nuclear magnetic resonance techniques, he made a sample of calcium fluoride containing approximately 10^{12} spins perform a quantum simulation of a spin system with long-range spin-flip interactions not known to occur in nature. Cory used this quantum simulator to perform a quantum calculation of the spin diffusion rate of the artificial system, obtaining a result compatible with semi-classical mean field approximations of this rate. To perform the same calculation on a classical computer would require a computer with at least $2^{\{10^{12}\}}$ elements, a number much greater than the number of elementary particles in the universe. Cory's technique is not amenable to simulation of arbitrary quantum systems, but a substantial set of systems can be addressed by these methods. Once 'general purpose' quantum computers can be constructed with more than 30 or so bits, calculations of the properties of a large variety of quantum systems should become more accurate on quantum computers than on classical devices.

6.2.4 Metrology and Few Photon Applications

The tools developed for Quantum Computation have great potential for several applications for measurement instruments. Qubits can be encoded in various physical systems: photons, ions, nuclear magnetic spins of molecules (NMR), quantum dots in semiconductors, etc. Not all these possibilities are suitable for a full quantum computer, because some seem impossible to scale up. However, they are all suitable for few qubits applications.

For example, the 2-photon source based on parametric down conversion used for tests of the Bell inequality (also for teleportation, and possibly for quantum cryptography) provide natural time and frequency bases: the 2 photons are created at the same time and the sum of their optical frequency is well defined. Hence, this type of source is a natural candidate for an instrument measuring chromatic dispersion, that is, measuring the time of flight in function of optical frequencies.

Highly accurate measurements of polarization mode dispersion have also been demonstrated (using type II down conversion, i.e. with the polarization of entangled photons). Yet another example, which was actually proposed first, is to use photon twins produced in parametric down conversion (the same photon pair source as used for tests of Bell inequality), to determine the efficiency of detectors without any reference to a calibrated source. Hence detection efficiency could be determined independently of black body radiation (i.e. independently of temperature). This theory has been known for many years, and experiments are now catching up.

Many standards, like the atomic clock frequency standards, rely on quantum physics. Hence the accuracy of many standards depends on our ability to manipulate quantum states. Consequently, the tools developed for QC have applications for improved standards. It has been suggested that the error correcting codes developed for quantum computing can be used to improve the accuracy of today's atomic clocks. The efforts to overcome and master environment-induced decoherence will lead to entirely new possibilities for sensors: the extreme fragility of entanglements between qubits can be used as a highly sensitive measurement of changes in the qubits' environment.

6.3 Conclusions: Measuring Progress

One possible benchmark for monitoring progress in this field could simply be the number of entangled qubits that researchers are able to manipulate. But a more significant challenge would be the industrial development of at least one application using the technology of QC in the next 5-10 years. Quantum cryptography is clearly one candidate. But it is equally possible that the first application will be in some measurement instruments.

If no application has emerged within the next 5-10 years, then it is likely that technological difficulties or a lack of imagination are providing a more serious obstacle than was first imagined. In both cases the field and its goals would require serious reappraisal.

Chapter 7
A Note on the Question of Scaling: Decoherence and Error Correction

Michael Brooks

Demonstrations of principle, such as a quantum gate with a single spin or ion or the realisation of entanglement in various physical situations, constitute first steps towards quantum information processing that require up- or down-scaling to achieve effective processing of quantum information. Decoherence makes increasingly difficult the up-scaling of quantum information systems to a few spins or ions. This process destroys the correlations among the individual spins or ions, losing quantum information into the environment and introducing errors in any QIP routine.

It is, perhaps, the main problem and challenge in scaling up to produce useful QIP systems. A primary goal is to understand how to avoid or combat the effects of decoherence.

Quantum error correction schemes represent a possible solution, whereby a single qubit is encoded by a string of qubits. The encoding is carried out such that an error in a single qubit will still allow the original qubit to be inferred from the code string and thus restored. Through the application of the powerful notions of quantum error correction, decoherence may be circumvented at the expense of a more complicated coding procedure using a larger field of qubits. The design of fault tolerant quantum systems must be a subject of continuing research if useful QIP systems are to be built; for example, research to understand properly the error limits that are imposed on the operations of quantum gates in order to be able to perform fault tolerant computation in practice.

Down-scaling quantum non-demolition (QND) measurements to the single photon level produces cascadable Controlled-NOT gates for quantum computation. However, at this level, the non-linear optical interaction that is at the root of QND becomes so weak that with present day technology it is no more effective. Cavity QED has been proposed in order to enhance it.

Theoretically, the prospect for coherent quantum computation is very good, provided that the decoherence satisfies some assumptions. These assumptions include:

1. that decoherence occurs independently on each of the qubits of the system

2. that the performance of gate operations on some qubits do not cause decoherence in other qubits of the system

3. that reliable quantum measurements can be made so that error detection can take place

4. that systematic errors in the unitary operations associated with quantum gates be made very small

If all these assumptions are satisfied, then we have a very strong result, one that says that fault-tolerant quantum computation is possible. That is, efficient, reliable quantum-coherent quantum computation of arbitrarily long duration is possible, even with faulty and decohering components. Thus, errors can be corrected faster than they occur, even if the error correction machinery is faulty. There is a threshold associated with this result; fault tolerance is possible if the ratio of the quantum-gate operation time to the decoherence time is smaller than some value (now estimated at around 10^{-4}), and if the size of the systematic error in the gates' unitary transformation is also no greater than 10^{-8} (possibly a pessimistic estimate).

Section II

Personal Perspectives

This section contains six presentations given by delegates to the Pathfinder Conference held in Helsinki in September 1998. They are conversational, not academic, in tone, providing personal insights into some of the implications of the quantum computing and communications research that is going on around the world. The content of this section also highlights some aspects of the field that are rarely discussed but may prove highly relevant to the development of this subject area.

Chapter 8
Solid State Quantum Computation: Prospects, Proposals, and Prejudices

Bruce Kane
University of New South Wales, Sydney, Australia

Although I am a solid state experimentalist, I have spent most of last year dispensing with experiments and thinking exclusively about how I could create devices that could have an impact on quantum computation. I have even stopped worrying about particular devices and have started thinking about the general principles that might be useful for guiding us in thinking about which particular systems and devices could have an impact in this area.

I'm not going discuss prospects, but prejudices. Basically, I'm going to discuss the assumptions underlying the proposal I have made in a Nature article, (Nature Vol. 393 p. 133 −14 May 1998). I think these are fruitful directions to take in solid state quantum computation.

The first prejudice I need to justify is: why solid state? The answer to that lies in the technology that has already been established for making solid state electronic devices. We might think that the quantum computer is going to look totally different to the classical computer, but it is legitimate to think that there's going to be a lot of commonality and a lot of similarity. The other thing to think about is that there's actually a $100 billion industry invested in manufacturing classical computers; if one hundredth of one per cent of it turns out to be relevant to quantum computation, that's a lot of money for everybody in the field. So maybe we should think about using some of the technology that goes into the Pentium processor to make a quantum computer!

Once you decide you're going to go with solid state, the next question is: what are the possible qubits you can use? We know about photons in cavities, electron orbital states, superconducting devices, electron spin states, nuclear spin states, but how do we decide where to go?

My second prejudice guiding me here is this: a quantum computer cannot be too coherent. I think we should concentrate on systems that have the longest relaxation times possible. If we go in the other direction and aim for systems that are going to be marginal, we're never going to get to the level where errors are less than 1 part in 10^4 - the level needed for error correction to beat decoherence. If we don't get that, then we will never have a hope of doing quantum computation. So let's look for systems that have very long relaxation times.

During my research I was very impressed to find that there are 1000-second electron spin relaxation times in a very familiar system: silicon. This was measured back in the 1950s by George Feher, who was doing electron-nuclear double resonance (ENDOR) measurements on doped silicon. Thanks to the weakness of spin-orbit coupling effects in silicon you can get very long relaxation times of the order of an hour at relatively low magnetic fields and low temperatures. The other advantage of Group IV as opposed to III-V semiconductors is that Groups III and V always have nuclear spins; there are no spin zero isotopes that are stable. However, in Group IV - silicon and germanium for example - there's at least one stable isotope that has zero spin; that means there are less degrees of freedom to decohere the system.

In response to this discovery, I fashioned a proposal for a quantum computer using this system, where we already have data. I chose phosphorus doped into silicon, where you can get one extra electron and one nuclear spin associated with each donor, and tried to design some kind of device - one that could be manufactured in the next few years - that could use this as a qubit.

Now I have to say that, although silicon is an ideal material, there's a legitimate reason to think that GaAs is going to be a possibility. The electron spin relaxation times in GaAs are quite long; possibly of the order of 100ms at low magnetic fields and at low temperatures. Strangely enough, the relaxation times of GaAs have not really been measured effectively - people are still disagreeing about what they are. There are also alternative systems like electrons on helium films, which have a certain overlap with the ion trap approach to quantum computation. So, while I'm going to focus on silicon, I by no means think that's the necessary conclusion you have to make.

Here I'll add in an extra couple of prejudices: first of all, it's OK to operate a quantum computer at low temperatures. Most people probably wouldn't think that's a spectacular statement, but, as a device physicist, if I ever set out to make a device that only works at a 100 millikelvin there's a lot of resistance. When we're considering quantum computation, however, I think that's legitimate - if someone can make a quantum computer that only works at 100 millikelvin, then you can be sure that people will still be interested.

The other prejudice has to do with which qubit you would like to use in phosphorus-doped silicon: electron or nuclear spins. I happen to think it's also a good idea if a quantum computation is slow. If you have a system with nanosecond dynamics and you have to get this one part in 10^4 accuracy with all your gates, then you're talking about sub-picosecond accuracy in your gates. On the other hand, if you have dynamics of tens of microseconds, it starts to become more experimentally feasible to have the necessary accuracy. So I've intentionally focussed on the nuclear spin, knowing that's going to create a computer that seems slow (I'm looking at 100 kHz clock speed), compared to what people would consider a state of the art computer.

I'm not going to get into the physics at all, except to show what we're proposing (Figure 8.1). Essentially it's something that looks a lot like a conventional metal oxide semiconductor structure, with the exception that what the gate will manipulate is not an electron current, but the position of the electron wavefunction. The electron surrounds the phosphorus donor as a kind of extended Bohr atom, and using an electric field you can move the electron around. By doing that in appropriate ways, you can do all the things that are necessary for quantum computation. You can do a selective NOT (flip a single spin in the computer), since the electrons control the Zeeman interactions of nuclear spin, and you can also do a controlled NOT by using an additional gate that sits between the donors and turning on and off the electron exchange between adjacent donors.

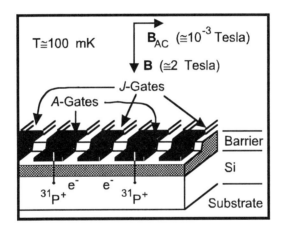

Figure 8.1: illustration of two cells in a 1D array containing ^{31}P donors and electrons in a Si host, separated by a barrier from metal gates on the surface. "*A*-gates" control the resonance frequency of the nuclear spin qubits, while "*J*-gates" control the electron-mediated coupling between adjacent nuclear spins. The ledge over which the gates cross localises the gate electric field in the vicinity of the donors.

There's only 100 angstroms separating the gates, and I put those phosphorus donors in very precise locations in the silicon substrate. Those are both things that you just can't go out and do: you can't make things at that scale and there's no way of putting donors into regular arrays. I'm going out on a limb here, but I think those are technologies where significant progress can be made in the next few years; certainly people are going to spend a lot of money to get to ten nanometer lithography and precision deposition. Doping is something that's also going well: there are certainly a lot of approaches that could be used to put dopants in regular arrays. So, while I realise these are impossible with current technology, it might well happen in the not too distant future.

The final prejudice I want to get into is the technique of single spin measurement. I should emphasise that nobody has measured a single nuclear spin, though there's lots of work going into it at the moment. My approach for some day making devices that can do that is to measure a single electron spin relying on the Pauli principle. Basically, it involves making a device that can tell you whether two electron spins are pointing in the same direction or opposite direction and looking for a detectable signal as a consequence of that. A picture of how that might work is in Figure 8.2.

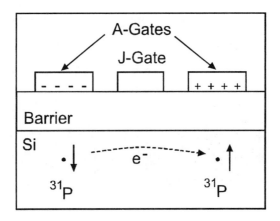

Fig. 8.2: only singlet electrons can make transitions into states in which electrons are bound to the same donor (D^- states). Electron current during these transitions is measurable using capacitive techniques, enabling the underlying spin states of the electrons and nuclei to be determined.

There would be two gates and two donors; you look for motion of an electron going between two donors. If both electrons are sitting on a the same donor (occupying the same orbital quantum state), the electrons must be in a singlet spin state, so if you can detect the electrons going into this state, for example by single electron tunnelling devices, you could measure spin.

Converting electron and nuclear spin is another challenge. Adiabatically converting the spin on one species to another is not that difficult if you have the precision control of the electron wavefunctions (I discuss this in the Nature paper); certainly, systems that use only electron spins are another strong possibility to look at, however.

So let's look at the implications of this area, irrespective of whether these particular ideas are valid. The first thing to look at is spin decoherence in solid state materials. In fact, we need to get solid state people talking to nuclear magnetic resonance people. These communities haven't really communicated in the last 35 years; I think there's a tremendous number of experiments waiting to be done just as soon as people start talking to each other. From the perspective of a device physicist - forgetting about quantum computation for a moment - developing devices that measure single spins is an extremely exciting challenge. Whether or not you can make lots of them into a quantum computer is a very difficult issue that we'll just have to put off for a while.

The other thing is, if we're going to go into this nanostructure regime, we're going to have to develop lithography/deposition almost at the scale of the atomic size - under 100 angstroms.

A lot of the resistance encountered in this field comes from people who say it's just playing around if the process is not applicable on 12 inch wafers. That's particularly true for the scanning tunnelling microscope people; they have lots of ingenious techniques for moving atoms around, but the people doing VLSI think it's impossible to imagine doing it over 12 inch wafers. This argument is not relevant for quantum computation - if you can make one quantum computer with an arrangement of 100 atoms that's really good news even if you can't possibly mass-produce the thing.

Also I really don't think the problem of scaling has been resolved. All the gates in my computer have to be custom biased in order to do quantum computation, but I can imagine doing custom biasing at the 10 to 100 qubit level. While I believe sound approaches for doing quantum computation at this scale are important, it's still much too early to say how we will get beyond this level to the size of machines necessary to usefully implement the algorithms of Shor and Grover, for example.

Chapter 9
Information is Physical, But Slippery

Rolf Landauer
IBM, Yorktown Heights

I am aiming to give the perspective of an outsider. I'm not really into quantum computing - in some ways, I'm a critic, although I am a friendly critic. To begin with the conclusion, I would like to point out that, in this field, people are always making comparisons with computing history, stating that quantum computing has reached this point, or that point. Currently, I believe, quantum computing is where Charles Babbage was: he had a relatively prophetic, far fetched and complete vision of what you could do in a mechanical way for handling data. He did not, however, have the remotest chance of actually doing it in his time. The real breakthrough came in 1890 or thereabouts when Herman Hollerith put together the right technology with the right application - tabulating US census data with electrical sensing of holes in punched card. That's the combination that started us on the whole road towards automated data processing.

I've been somewhat critical of quantum computing, but it's not just this field - I've also been critical of a great many other adventurous suggestions for how we can replace the transistor. Many of these technologies have serious problems, many are not totally convincing in certain ways. Many of them suffer from the fact that they don't do what we do in ordinary CMOS logic: they don't reset the signal at every stage to the desired value of 0 and 1, so the signal can drift away. All-Optical logic, for example (I'm going to use this as an instructive case history), was said to be 'bound to come' by the chairman of the University Grants Committee in 1984 - a distinguished member of the British academic establishment. He listed milestones in computing, a non-trivial fraction of which have to do with optics. In the end, he said, All-Optical computing is 'bound to come'. Such optimism is expected from those asking for money, but rarely from those who give it out!

But in a recent paper ['Perspectives in Optical Computing', Computer, February 1998, pp22-25] John Caulfield gives an unprecedented example of honesty:

Thirty years ago, this is what I knew for a fact: electronics is reaching its natural, technical limits. Optics alone can offer massive parallelism...

...I was all wrong. As optical practitioners learned more about the nitty-gritty of 'real' computing (the computer science stuff about complexity, conditioning, accuracy, and so on), fear set in. The real problems were much harder than we thought. Added to that, electronics also kept ignoring our prophecies of slowdown. Even in optics' most natural domain, Fourier Transforms, electronic chips began to rival optical ones in throughput and destroy them in accuracy...

...A huge effort has gone into what some optical practitioners have viewed as a competition between optics and electronics. This has been a tragic waste of time, money and talent...

... Those of you in electronics have tried for decades to tell the optical community these things, but we did not want to listen. Neither you nor the electronics are our enemies. We are.

It's praiseworthy that someone should be so honest. Now what do we learn from this?

All-Optical computing had many problems, most of which were written down in the literature before it became fashionable or was given any attention. I want to go through one of them because I think this community can learn something from it. All-Optical computing was tied to the wavelength of light. That is large compared to what we can do with a transistor - therefore the propagation distances are tied to the wavelength of light. Now All-Optical computing had a choice of transmission means: optical fibres, guided optics replacing transmission line, or a free space propagation, hopefully getting most of the light where its supposed to go. In either case, the interconnection distances become very large compared to what we can do with the computer. The propagation delays will kill us - it's hopeless.

Why is this important? It's important for the quantum computing community because it says that to evaluate a technology, you cannot just think of the gate and how it works, you have to understand the whole structure in real space.

The typical advertisement for quantum computation is that transistor components will get smaller and smaller anyway, and quantum effects are inevitable, so we might as well make good use of them. There are two things wrong with that advertisement. First of all, the logic is wrong. Just because some day the gate will become thin enough for the carrier to tunnel through it, and that's going to be a problem, that doesn't mean that's an easy way towards handling information in a totally quantum mechanical way.

But aside from that logic, the notion that components are getting smaller and smaller isn't so clear: the reality is that most people in the semiconductor industry are beginning to admit that miniaturisation is slowing down; Moore's

Law does not hold forever. It is slowing down not for physics reasons, but for economic reasons. That, of course, doesn't mean that the semiconductor industry is going to remain static: people working in quantum computation are competing with a moving target! There are other ways in which semiconductor components can be improved. It's my guess that the likely course of the transistor/semiconductor business is that attention will be diverted from miniaturisation towards other directions: three dimensional integration, for example, could become a serious objective. After all, it's the density of the components, not the device size, that counts.

The complexity of modern computing comes largely from its packaging hierarchy. Chips are on cards, cards are on boards - that's where all the expense and the space and the money come from. And there is room for cleverness there - as a playful example, one could define a high speed optical bus that snakes past a lot of chips and lets them talk to each other, getting rid of a lot of junk.

I've admitted that miniaturisation is likely to be slowing down, on the other hand, I have to warn you that in my files at home I have many premature obituaries for the transistor . Of course everything has limits - CMOS and the transistor will have their limits - but it has often been predicted prematurely.

Now, a word about error handling. The quantum computation community has made remarkable progress, far more than I allowed for some years ago. However, there are still problems. The community understands decoherence very well, but, perhaps a little less honest and a little less clearly faced are flaws in the Hamiltonian. They come up for two reasons.

For a start, no manufacturing process is perfect, so the bits are not going to be spaced exactly as desired. But there's also a more serious problem. In the typical proposal that's been made - actually executing quantum computation - those physical implementations typically provide a very good approximation for the Hamiltonian or the unitary transform you want. But they don't carry it out exactly. The basic operation is not exactly what is needed, so every single gate in the device is slightly, and systematically, off what is wanted. Maybe the error control procedures suggested can handle it, but it hasn't been honestly discussed in detail. Very few papers actually discuss this problem, rather than claim that the manufacturing flaws are going to be statistically random. If all the components are slightly off, how does the machine - which, by the way hasn't read the papers and doesn't know how it is *meant* to behave - know what to do?

There is also unintended coupling - we don't have a neat set of gates as is drawn in all these diagrams. Some of the components which shouldn't be talking to each other are unintentionally coupling. Just as with error control, once the problems are admitted, there may be clever ways of handling them, but first they have to be admitted and discussed. Error control methods always handle a limited error space - no error control method we've heard about will, for instance, protect the computer from being swallowed up by an earthquake. In the papers it is always discussed in very analytical terms - not in discussions of actually how much variation in space and bits I am allowed. I'd like to read about a scheme that deals with these things.

In fact, the thrust of much of what has been discussed in quantum computation is pet inventions. People discuss them and ask what they might be able to do in 5 or 10 years, and what resources they need to achieve this. I think that way of talking suppresses some important questions: there is danger in premature concentration. There are various issues that need to be discussed. One of them - my personal favourite - is the issue that information is physical, and that means it will allow certain possibilities. This community has taught the world that we can do something with quantum computation. Quantum superpositions or information-bearing states are useful and interesting and we have to learn what to do with these things. They are possibilities, but there are also restrictions: we live in a finite world, we don't have unlimited memory, we don't have unlimited Turing machines. It's not available in principle, not just within your budget. So, we ask, what is the impact of that on all information handling?

We've all learned the mathematics, we've been taught there are certain operations, and now we have to come back and ask if these are executable functions, for instance. I think there are important questions, that get suppressed by premature pitching of what we might be able to do in five or ten years with various approaches. I would like to merely encourage the study of questions such as these.

Further Reading

R. Landauer, 'Need for Critical Assessment', IEEE Transactions on Electron Devices, vol. 43, No. 10, October 1996.
R. Landauer, 'Information is Inevitably Physical', submitted to *Feynman Lectures on Computation*, vol. 2, edited by A.J.G. Hey (Addison Wesley, Reading, 1998).
R. Landauer, 'Supposed Universality of 1/3 Shot Noise Reduction' (contains an appendix evaluating adventurous device proposals), submitted to *Microelectronic Engineering* as part of the Proceedings of *New Phenomena in Mesoscopic Structures*, Hawaii, December 1998.

Chapter 10
Nanocircuitry, Defect Tolerance and Quantum Computing: Architectural and Manufacturing Considerations

R. Stanley Williams
Quantum Structures Research Initiative, Hewlett-Packard Laboratories

One of my titles is 'Director of Basic Research' for Hewlett Packard Laboratories: it's my privilege to conjecture what long-term research will benefit Hewlett Packard. I hire people, obtain resources for them and set rather loose boundaries for their research areas. I am a big fan of quantum information; I think it's going to be very important scientifically and technologically, but I don't really know what specific area to invest in now. I haven't hired anyone to work directly on quantum information in my lab, even though HP as a company has a roughly $3.1 billion R&D portfolio devoted primarily to information technologies.

Why does HP have such a small effort in quantum information? Why don't we put together a program to become the world's best center in quantum computation?

The answer is that even in the research labs we have to be 'cold blooded businessmen'. We exist in an extraordinarily competitive environment. Probably only half the companies now working in electronics will survive the next decade, and we intend to be one of them. But in order to do so we have to invest our resources very wisely; we are responsible to our shareholders.

To show our reasoning, I will go through a toy calculation of the kind we might do when deciding whether to fund any project.

The first question is this: what is going to be the total world market for the technology - chips or chip equivalents - that will be available? The answer is, looking 15 years ahead, $1 trillion per year (in dollars of the day) for electronic devices. Today the market is about $200 billion.

We then have to ask: under the most optimistic scenario possible, what fraction of this market will belong to quantum computation? Being extraordinarily kind, let's say it will hold ten per cent. That would give us a market share of $100 billion. Now, how much could HP capture if they went after it very aggressively? HP is a $50 billion per year company, so it wouldn't be out of line to assume HP could capture ten per cent of the market. So let's say we could sell $10 billion per year worth of quantum computation equipment.

Now the question is, if we could sell that 15 years from now, what is the appropriate level of investment today for that income stream? The first thing we ask in response to this is 'what's it going to cost to develop all this stuff?'. Historically, we know that development costs are about ten per cent of sales per year for an advanced technology. If you're going to be selling quantum computation equipment, then this is what you've got to spend per year on development to keep your edge: $1 billion per year.

How much of that is research? Research is about 10% of development, so an appropriate research investment in quantum computation would be something like $100 million per year, but that's talking about a market that's 15 years down the road. In the industrial area, you have to pay for money, and within American companies that cost is about 15 per cent per year, so to discount an investment 15 years into the future requires dividing by another factor of ten: now we are down to $10 million per year.

But there's still one more thing: the risk involved. Fewer than one in three technologies developed in research labs go through to manufacturing. That's a 33% risk factor - organizational risk. Even if you make it to product, it might not sell – market risk. On top of that is the technological risk that the idea doesn't even work to begin with. HP Labs may be willing to take on a project if we go through a risk analysis and decide the total chance of success is one in ten. That's actually better than many venture capitalists can do but it also requires another discount by a factor of ten for the investment to make sense in terms of risk.

That calculation leaves us understanding that, even when addressing a significant share of a $100 billion market that is 15 years in the future, the amount of money we should be spending now is about a million dollars per year. In an industrial laboratory environment that's about three researchers with their respective overhead costs. A company that intends to be involved in the computational business a couple of decades from now should definitely be looking at quantum computation, but it is betraying its stockholders if it is currently spending much more than a million dollars per year working on this area. Given the different types of internal and external quantum information research we support at HP, we are within that limit.

This is an important lesson in the economics of doing research in the industrial arena. Let's look at the total global R&D effort in electronics for the 1997 calendar year (source – October 1998 R&D magazine): Siemens, investing the most on research and development in electronics last year, spent $4.6 billion. Next come IBM with $4.3 billion, Lucent Technologies with $4.1 billion, Hitachi with $3.4 billion, and HP with $3.1 billion. Worldwide, there are another 25 companies that spend $1 billion or more on electronics R&D (10 American, 8 Japanese, 7 rest of the world).

Market Size	→ $1,000,000,000,000
Market Share	→ $100,000,000,000
Sales	→ $10,000,000,000
Development Costs	→ $1,000,000,000
Research Cost	→ $100,000,000
Money Costs (0.85^{15})	→ $10,000,000
Risk (technical, market, organizational)	→ $1,000,000

Figure 10.1: rationale behind Hewlett Packard's risk-weighted investment approach.

Thus, there are probably only about 30 companies in the world large enough to justify a $1 million per year investment in quantum information, and many of them will probably not do so. Thus, until the risk levels decrease, government expenditures on quantum information R&D will be significantly larger than those made by industry.

Given all this information, what can I point at as a reasonable strategy for us to follow?

At Hewlett-Packard Labs, our long-term, high-risk project addresses almost the entire electronics market for the 2010 time frame. In fact, that is an essential target: if you are doing fundamental research in an industrial lab, and you're not trying to do something that will completely change an entire market, you're not addressing the needs of the company.

In my lab, we're looking at economics, chemistry and physics to see if the cycle of history will repeat. Look at ENIAC, the Electronic Numeric Integrator And Calculator built in 1946 by the Moore School of Engineering of the University of Pennsylvania, for example. ENIAC and the Intel 4004 microprocessor have essentially the same computational capacity, but are totally different in size and in power consumption. ENIAC required 100 thousand times more electrical power than the Intel chip to do the same amount of computation in the same amount of time. There's still a factor of 100 million improvement in the power consumption of computing devices over what we have today before we reach the limits of nonreversible circuits. To quote Feynman: There is plenty of room at the bottom. From where electronics stands today, there's great opportunity for improvement without having to involve reversible or quantum computation.

Is there another 'transistor' that's going to arise, taking us from the integrated circuit paradigm we're in now to something so different that it justifies doing fundamental scientific research in an industrial lab?

My trivial answer to this question is that in 1938 the transistor hadn't been invented, and yet the people at Bell Labs were hiring the best solid state physicists and chemists they could find, and motivating them to invent a solid-state switch. They knew it would be better than a vacuum tube, but they had no idea how much better: there was no conception of just how good a solid-state switch could be.

We think we can do at least 100 million times better than can be done today using a quantum state switch - something that would operate at the nanometer scale, and wouldn't require superposition or entanglement.

Many times in the past people have come up with something: GaAs - 'the semiconductor of the future that's going to wipe silicon off the map' - or superconducting electronic circuits, or something else. Many people have come along with adventurous proposals, as Rolf Landauer calls them, and have been smashed by the silicon steamroller.

Silicon has been following an exponential growth in its capabilities and it has been doing so for over 30 years. There's even a very detailed roadmap, published by the Semiconductor Industry Association of the United States, which plots what needs to happen until 2012 to keep this growth going. Short term, now that everyone has a roadmap, the managers of the fabrication facilities at companies are telling their engineers to beat it, and so progress in developing further generations of electronic circuitry is actually accelerating.

The increase in the number of transistors on a chip has been documented and its progress is discussed as 'Moore's Law'. But there is another factor to consider: the cost of building a fabrication plant has also exploded. In 1995 fabrication plants cost $1.5 billion each. Worldwide, at the beginning of 1995, work was begun on 70 fabrication plants, and 30 of those never produced a single chip. That overcapacity issue is something we're suffering from right now: from 1965 to 1995 the cost of a fabrication plant had doubled every chip generation - approximately 3 years - so in 1998 fabrication plants are $3 billion each. I've extrapolated this data out into the future, and I make it $50 billion for a single fabrication plant in 2010. No company and very few countries could afford to build such a factory.

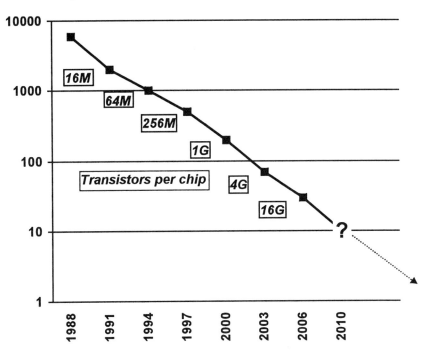

Figure 10.2: the declining trend in the number of electrons on a transistor as time progresses. Also shown is the growing number of transistors per chip.

Economically, the system cannot support this growth in the cost of factories. At the beginning of this year, 1998, the construction of about 20 fabrication plants was started. Before the first devices have come out of any of them, at least four have been closed. Much talked about physics issues, revolving around the problems with shrinking of transistors, will probably not become a problem; economic issues are going to slow Moore's Law before physics, unless a new manufacturing paradigm is found.

So, given that there may be a window of opportunity to come in and replace integrated circuits, what other opportunities are there?

To research this, I decided to learn a lot more about computers and computer architecture. For over a year we had detailed discussion sessions involving how computer architecture, chemistry and physics could work together. It took six months before we were speaking the same language, where we could

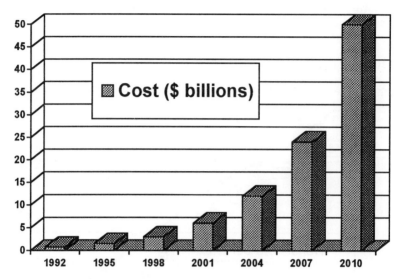

Figure 10.3: the increasing cost of chip fabrication facilities. The cost is rising much faster than industry sales.

say words that roughly registered equivalent meanings in the different participants.

The computer architects had made a machine that they called 'Teramac': it performs 10^{12} operations per second, and it is a Multi-Architecture Computer. The detail is given in Science, 12 June 98, Volume 2a 0, pp. 1716-1721. It was based on standard silicon integrated circuit technology. It was quite large, of refrigerator size - it weighed about 400 pounds. It had the equivalent of 256 processors, but the logic operations were performed with look-up tables, i.e. memory, rather than standard processors.

Teramac contained 220,000 mistakes - in effect, it was built out of defective parts. Its inventors designed the chips they wanted, sent the design to the fabrication plant, got costings for so many good chips and took the defective ones too - in this architecture the defective chips were every bit as usable as the non-defective ones. They did the same with several other components: in the end they had 800 chips, miles of wire, and the entire system was assembled as quickly and cheaply as possible. Thus, there were lots of manufacturing mistakes in both the components and the way they were connected. Even though it had a clock speed of only 1Mhz, for some applications it managed to outperform - by a factor of 100 - the best workstations available at the time, which operated at more than 100 MHz clock speed.

The Teramac team was surprised how quick and efficient it could be. The communication network put in to handle the defects had also made the computer much more efficient at doing some things than the designers had expected. One issue: how long does it take to compile a program on this computer? Given arbitrary logical and physical configurations, compilation could be an NP hard problem. However, it turned out that the compilation time was linear with the size of the logical system being configured onto the hardware.

Why doesn't HP sell these computers? These technical advances came at a cost of additional silicon area: all these chips added up to a quarter of a square meter of silicon. About 90 per cent of that area went into the communications network that enabled the defect tolerance, not into the lookup tables. Thus, given current Si technology, a Teramac-like machine cannot compete economically with standard computer architectures, in which the chips are all perfect.

So it's not economically viable to sell something like this using today's technology. But given Moore's 2nd law - the one about the cost of fabrication plants - we think there's eventually going to be a crossover to this type of architecture. We've come up with a quantum state switch that we think is going to work in this architecture, and we have a model process to assemble it. Conceptually, we plan to make device components that recognize each other chemically, so, by properly bringing the components together, they will assemble themselves into a circuit. The random errors or mistakes introduced into such a self-assembled system should be compensated for by the proper defect-tolerant architecture.

Chapter 11
Quantum Computing and NMR

Jonathan A. Jones
University of Oxford

Let's start off by looking at what we require to implement quantum computing. Any quantum computer needs four things:

1. Qubits
2. An adequate set of gates
3. An initialisation operation ('CLEAR')
4. A readout mechanism

The issue facing anyone who wants to make one, though, is how easily you can do this. In nuclear magnetic resonance the first two are very easy; nuclear magnetic resonance is the study of spin half particles, and these make good qubits. Gates are also incredibly simple - so simple that we've been doing two qubit logic gates with NMR for over 30 years (but we only found that out a few years ago). Bigger operations are, of course, more tricky, but these 2 qubit gates are genuinely trivial.

On the other hand, these other operations - initialisation and readout - are rather more interesting. Initialisation is very difficult: in NMR it can be done, as was shown a couple of years ago by a group in the United States - the group that set off NMR quantum computing (see the historical resume of NMR at the end of this chapter). It is, however, inefficient. In fact it is exponentially inefficient, which means every time a qubit is added the efficiency goes down by a factor of two. There's also big problems because NMR quantum computing is essentially a 'once for all' operation; at the start the system is initialised and then the whole calculation has to be run. The qubits can't be selectively reset to 0 or 1, which most other techniques can do - or claim to be able to do. That's a big problem, particularly with error correction. Error correction becomes very difficult to do if you can't reuse your qubits.

> **Qubits**
>
> The two Zeeman levels of a spin-$\frac{1}{2}$ nucleus in a magnetic field provide a natural implementation for NMR quantum computing. Obvious candidates are ^{1}H, ^{13}C, ^{15}N, ^{19}F, ^{31}P.
>
> $H = -\gamma\hbar B I_Z$
> $\quad = -\omega I_Z$
>
> ω is called the Larmor frequency. For ^{1}H nuclei in "typical" NMR magnetic fields the Larmor frequency is about 500 MHz, corresponding to RF.
> Spin-0 nuclei can be completely ignored. Nuclei with spin $> \frac{1}{2}$ can be largely ignored.

Readout is a bit of a strange concept in NMR, because we have a macroscopic ensemble. That has obvious consequences; in terms of physics it means that everything one measures is an ensemble average, rather than a single system state. On one level that doesn't really matter because everything we want to know can be translated somehow into an ensemble average: one can always find the right operator to do it. But it does require that some thought goes into what is going on. It also has a huge advantage, however, because NMR observation is essentially non-destructive. This upsets the quantum theorists who say that if one observes a quantum system it is 'broken': it isn't true in NMR; at least it isn't true in practice. The reason for that is simply that we have a very large number of quantum systems, and indeed identical quantum systems. When one observe such ensembles, you know those observations are essentially passive, and therefore you don't break the system.

Just to give an idea of the experimental details, I will outline some of what goes on in an NMR experiment. A typical 'high resolution' NMR sample comprises a few milligrams of some organic compound dissolved in an appropriate deuterated solvent. The normal sample size is about 0.5 ml.

For our NMR quantum computer, we use a solution of cytosine in deuterated water. This gives a two spin system (two ^{1}H nuclei), suitable for a two bit quantum computer. Another popular choice is ^{13}C labelled chloroform: this provides another two spin system, one ^{1}H nucleus and one ^{13}C.

The sample is placed in a strong magnetic field, typically 10-12 Tesla, resulting in a ^{1}H Larmor frequency of 400-800MHz. Superconducting magnets are almost always used. In order to observe the natural linewidth of the NMR signals, the field must be held constant and homogenous to about one part in 10^{10}. This is achieved through a variety of small electromagnets. The sample is surrounded by one or more high Q tuned RF coils, which are used to apply RF pulses to the system and to detect the resulting signal. A typical RF pulse lasts about 2-20μs.

Gates

Various universal sets of gates exist in principle. In practice the most convenient set is:

1) A complete set of 1 qubit gates (any arbitrary rotation in the Hilbert subspace corresponding to a single spin).

In NMR this is easy as resonant RF pulses provide a complete set of 1 qubit gates.
We can address each qubit individually as different spins have *slightly* different Larmor frequencies:

$$H = -\gamma\hbar B_{local} I_Z$$
$$= -\gamma\hbar B (1-\sigma_I) I_Z$$
$$= -\omega_I I_Z$$

The shielding constant, σ_I, depends on the local chemical surroundings of the nucleus (the chemical shift).
For ^1H nuclei σ varies by about ten parts in a million (10 ppm), so the Larmor frequency varies by about 5000 Hz.

2) Any one non-trivial two qubit gate (any non-trivial spin-spin interaction).

In NMR spin-spin coupling (J-coupling) provides a route to a large number of two qubit gates

$$H_{J(IS)} = 2\pi J_{IS} I \cdot S$$
$$\approx 2\pi J_{IS} I_Z S_Z$$

(the weak coupling limit, i.e., first order perturbation theory).

This is *not* simple dipole-dipole coupling, but a hyperfine effect, related to the Fermi contact interaction, and transmitted by valence electrons.

So, where are our limits? There are a lot of them, and this is where we stop feeling smug. The limit everyone always talks about is the exponential inefficiency in the preparation of pure states. It is something to worry about, but it's not as bad as everyone makes out. Currently with NMR, not really trying - just

---

CLEAR

CLEAR is the operation which sets all qubits equal to 0. CLEAR is non-unitary, and so must be achieved by a non-computational process. Typically it is achieved by cooling the system to its ground state, but in NMR the energy gaps are so small compared with thermal energies that this doesn't work at any sensible temperature. Furthermore an NMR sample contains not just one single molecule, but a very large number, effectively a statistical ensemble. Any NMR observable is averaged over the ensemble of spins. Thus an NMR system must be described by a density operator rather than a ket.

In fact this effect comes to our rescue! In an NMR sample there will be a very slight excess (typically about a few parts in a million) of spins in the low energy state. All NMR observables are traceless, and so the vast majority of spins make no net contribution to the NMR signal: only the small excess is seen.

The situation is much more complex when dealing with two or more qubits, as it is necessary to observe only molecules where *all* the spins are in the ground state. In 1996 Cory *et al.* showed how it is possible to use NMR techniques to distil an effective pure ground state from the equilibrium Boltzman distribution. This process is exponentially inefficient, and so limits current NMR quantum computers to about 20 qubits. Various other approaches have been described since then; all suffer from similar problems.

---

doing things the obvious way - it is limited to about out 10-20 qubits. You can extend that, with optical pumping, cooling things down, and all sorts of other tricks, maybe up to 30 qubits if one is really good at it. Unfortunately, that's completely irrelevant because other things are going to kill us first.

Decoherence is a real problem issue. It's one that everyone in the field shares and understands, and we all have very similar problems; I'm interested to note that most techniques seem to have similar decoherence/gate time ratios, with just a few exceptions. The effects of decoherence are clearly visible in our experiment - we can see them quite accurately. One might not think that from the early NMR papers, but in fact they were being economical with the truth. They took the longest decoherence times they could find - about 3 hours - and then they took the shortest gate times they could - 100 microseconds. Then they put them together, got a very good ratio, and ignored the fact that gate mechanisms are decoherence mechanisms: if you have fast gates you have fast decoherence. One can't just pick systems at random like that - you have to be honest. I think that, in reality, it's going to limit us to 5 - 10 qubits.

> **Readout**
>
> Readout requires some method of determining the state of one or more qubits. In principle this is easy - just measure the NMR spectrum of the corresponding spin. However, the ensemble nature of NMR quantum computers can complicate matters. For algorithms which end up in a pseudo-pure state this isn't an issue. For algorithms which end up in superpositions it can be an advantage, a disadvantage, or an irrelevance.
>
> Note that to a good approximation NMR observations are *weak*, that is they do not perturb the spin system! Thus there is *no* equivalent to projecting the spin system onto a basis of eigenstates. Weak measurements are fairly unusual in quantum mechanics, as they are confined to studies of macroscopic ensembles of identical quantum systems. This gives rise to some unusual effects in NMR and NMR QC:
>
> 1) Normal NMR detection comprises the *simultaneous* and *continuous* observation of the non-commuting observables, I_x and I_y!
>
> 2) The state of an NMR quantum computer can in principle be monitored *during* a calculation!

The next thing I have to mention is the 'house of cards' effect: life gets difficult as you build bigger and bigger systems. People say things are scalable, and we know that some things are scaleable in principle, but others are not. But one has to consider even those things which are scalable in principle - are they really scaleable in practice? This is a question we have to think about very carefully. With some of the obvious NMR implementations, the argument applies very badly because the complexity scales exponentially with the number of qubits - and that's a real killer.

On the other hand, there are slightly more subtle implementations, in most of which you abandon the ability to have any spin talking to any other spin. Essentially, one uses a linear chain of spins - we know we can do that. It actually works with only a polynomial overhead, and if you do that then life gets more difficult roughly linearly, or that's the theory.

In practice, we have to think about where life is going to get difficult. It's going to get very hard after about five qubits; to jump beyond six qubits is going to be very tricky.

This is for technical reasons which have to do with the way spectrometers are at the moment. It is currently difficult to address more than two frequencies per channel using the obvious techniques. There are ways round it, but it seems that the obvious techniques won't work beyond six qubits. Lastly, there's a problem which no-one talks about, which may turn out to be a real killer. That problem is selective excitation.

We know that we can't actually address each qubit individually by where they are in space. This is what everybody else does - it's a lovely way to do things - but, unfortunately, because we're dealing with an ensemble, and we're also dealing with a liquid, we can't. Instead we have to find another way of addressing them: we use frequency selection. In principle that's a great technique, but unfortunately the spectral bandwidth available for us to work in is depressingly narrow - you can't fit that many qubits into it. For a given nucleus, more than 5 or 10 will be difficult.

So, when we put all these things together, the number that floats around is 5-10 qubits. That's what I think we can do without much real difficulty, but beyond that it is going to start getting very hard.

To summarise, let's take a quick look at the future. NMR, I would claim, is an excellent toy system; probably the best toy system we know at the moment that does quantum computing. It's very well developed, especially since the theory of how to use spin half NMR has been done for thirty years. The necessary engineering has been done, and its commercially available.

I think its a great discipline and I think we're going to be in the lead for about 5-10 years before the other proposals start catching up with us. On the other hand, we do have real problems facing us after that period: life's going to be very easy for a while, and then we're going to hit some rather nasty barriers. After that I don't know where we're going to go. I don't think you can seriously propose that NMR is any sort of route to a general purpose quantum computer, but there's a great deal that we can do in the mean time. It's also going to teach us an awful lot about physics.

NMR Quantum Computing: What can we do so far?

1996   Cory *et al.* Demonstrate pseudo-pure states and simple logic gates.

1996   Chuang et al. Discuss (but do not demonstrate) similar approaches.

1997   Jones & Mosca demonstrate Deutsch's algorithm on a two qubit NMR QC based on cytosine.

1997   Cory *et al.* demonstrate 3 qubit error correction.

1997   Chuang *et al.* demonstrate Deutsch's algorithm on a two qubit NMR QC based on chloroform.

1997   Cory *et al.* demonstrate 3 qubit error correction.

1997   Chuang *et al.* demonstrate Grover's quantum search algorithm.

1998   Jones *et al.* demonstrate Grover's quantum search algorithm

1998   Linden *et al.* demonstrate a three qubit Deutsch-Jozsa algorithm

1998   Jones *et al.* demonstrate approximate quantum counting with two qubits.

The Future:
- NMR provides an excellent 'toy' system for investigating quantum computing.
- It is likely to remain in the lead for quite a while.
- The ensemble average nature of NMR QC gives it some unique advantages over more conventional approaches.
- However NMR is very unlikely to provide a route to a general purpose quantum computer.
- There are some extremely interesting speculative proposals to combine NMR with solid state techniques which might just work.

Limits to NMR QC

Exponential inefficiency in preparation of pseudo-pure states. This is the most widely discussed limit, but probably the least relevant! It will limit NMR QC to about 10-20 qubits. Various techniques, like optical pumping, might extend this a bit.

Decoherence. The effects of decoherence are clearly visible in our two qubit implementations.
The figures quoted in early papers are extremely unrealistic!
The problems are likely to become serious with more complex systems, and might limit NMR QC to about 5-10 qubits.

Pulse sequence complexity. With increasing numbers of spins the pulse sequences required to implement gates will become more complex. With systems where all the spins are coupled to one another (so all possible two qubit gates can be implemented) this complexity will grow exponentially.
With 'linear systems' where the spins are only coupled to near neighbours, the complexity will only grow linearly. But the polynomial overhead involved will make the effects of decoherence more serious; life will get difficult after about 5 qubits.

Selective excitation problems. Addressing individual qubits is based on their individual resonance frequencies, but the frequency space available is small. There is only room for, say, 5-10 qubits in the ^{1}H spectral width. These problems could be reduced by using different nuclei; but we will eventually run out of nuclei.

In practice the limit for NMR QC using the current approach is likely to be about 5-10 qubits.

Chapter 12
Quantum Networks and Quantum Algorithms

Vlatko Vedral
University of Oxford

I plan to explain how to do simple arithmetic operations, and then I want to demonstrate a simple example where we can easily see why quantum computing is more efficient than its classical counterpart. Just to remind you about looking at gates, Figure 12.1 is a simple network that accomplishes addition. Now this cannot be done on a quantum computer in this way, and there's an easy way to see why that is true. Take the first gate as an example: we don't even need to know what the gate does, what we need to see is that there are two inputs and one output. Therefore quantum information is lost on its way through the gate, and so this cannot be done reversibly. Since quantum computation is governed by unitary transformations, you need to do this reversibly. You simply would not be able to implement the first gate in Figure 12.1 on a quantum computer. The rule of thumb for doing this reversibly is that one has as many qubits coming into the gate as there are coming out of the gate. Let's look at a very simple example: addends in this case.

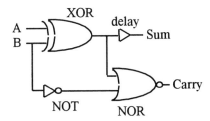

Figure 12.1: an addition network. This is irreversible, and would therefore be impossible to implement in a quantum computation.

The left hand side of Figure 12.2 shows what is needed for basic addends. There are four qubits in the input, and there are four qubits in the output. This is just a basic network for calculating carry in the additions. You add three simple bits, you quote the result a plus b, and the new carry is computed. In this case just using a Controlled-Not gate and a Toffoli gate twice.

Figure 12.2: carry (left) and sum networks.
c - carry; a, b addends; c' - new carry; r - result.

Another thing one needs is the network for computing the sum, shown on the right hand side of Figure 12.2. There is a carry, and there are two numbers to add - a and b - and out comes the result.

As an aside, I should point out that there is nothing intrinsically quantum in this network; this is actually a network stolen from classical computation. But in quantum mechanics the difference is that the input can be in a superposition of states. That is worth bearing in mind.

Just to show you how the addition can be done now we go to the plain adder (Figure 12.3). There are two registers as input - two numbers a and b - and you have output: a and a plus b. You can also see certain bits here, the carry bits, which are initially set to zero. a and b are just digits, and the boxes are the gates representing the carry and sum from the previous figures. Now we need to focus on two things. Firstly, this network has a peculiar 'V' shape - I will return to this in a moment. Secondly, this network is entirely reversible, so if my input is a and b, and the operation is a and a plus b, I can also have subtraction if I run the network backwards - by making addition reversible, I also get subtraction.

Let's now look at how to do basic simple arithmetic using this network to see just how simple it is. First, we'll deal with multiplication. Once I have addition I immediately have multiplication: I merely perform addition a number of times. X multiplied by Y is simply X plus X plus X...., repeated Y times; I just repeat my previous network. If I want to exponentiate, which is a key element of Shor's algorithm, it's just as simple. A to the power of X is A times A times A..., repeated X times. Now we have all the basic arithmetic we need for a quantum computer.

But you might also have noticed that there were some extra bits in the addition network. The carry bit, that was set initially to zero, was also zero at the end. These were auxiliary bits, and I didn't really need them at the end of the computation. We might ask ourselves, if we repeat one operation many times the way that it shows here, will we not generate a lot of unnecessary rubbish at the

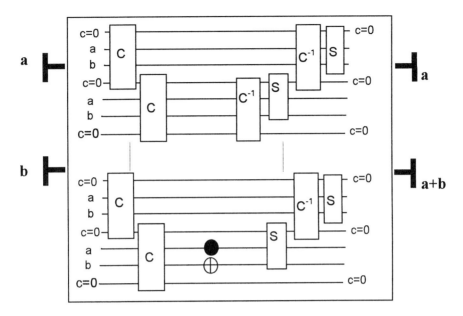

Figure 12.3: the Plain Adder. a and b could be strings of binary digits (e.g. a=0101001, b=101100); a and b within the Adder are distinct bits in the binary decomposition of those respective strings.

end of the computation? If this is not really necessary can we somehow avoid it? We can. This is another trick stolen from reversible computation. It is exactly the same in quantum computing - I call it 'Garbage Disposal'. We have two input registers: one is x and the other is set to zero. What we can do is apply a gate and get x and f(x) in the second register - this is a basic operation: it could be addition, it could be exponentiation or anything. Suppose now we want to compute $f^4(x)$ - equivalent to f(f(f(f(x)))) - which is more complex. That means we want to repeat this operation four times - this is exactly what is needed for multiplication if we know how to add.

Now we have prepared five registers. The first contains x and the rest are just set in state zero. What we do now is apply the basic gate to the first two registers and we get x and f(x). The rest are unchanged. Now apply the same trick to the second and third register, and we get f(f(x)) in the third register. We can iterate this procedure and get the result at the end.

Now we can see that the first register contains x and the last register contains what we want, but we don't need the middle three registers at all. So if we were to continue with this, we would generate unnecessary garbage. In reversible computing this might be garbage, but in quantum computing this is like our environment, which causes our register to decohere. That happens because, if we trace this out, it will be the other two registers in a mixed state; it is very important to get rid of this garbage in order to control it. So how can we reset this to zero?

$f^4(x) \equiv f(f(f(f(x))))$

We know how to perform:

$$|x\rangle |0\rangle \rightarrow |x\rangle |f(x)\rangle$$

Therefore we perform:

$$|x\rangle |0\rangle |0\rangle |0\rangle |0\rangle \rightarrow |x\rangle |f(x)\rangle |0\rangle |0\rangle |0\rangle$$
$$\rightarrow |x\rangle |f(x)\rangle |f(f(x))\rangle |0\rangle |0\rangle$$
$$\ldots$$
$$\ldots$$
$$\rightarrow |x\rangle |f(x)\rangle |f^2(x)\rangle |f^3(x)\rangle |f^4(x)\rangle$$

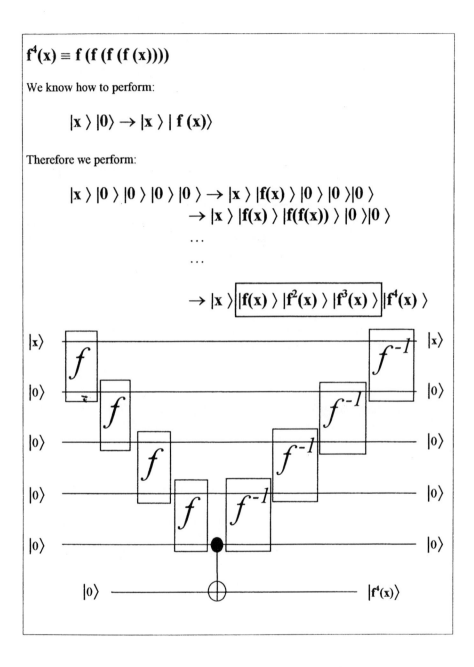

Figure 12.4: the process and network for 'Garbage Disposal'

It's actually very simple. We copy the result to yet another register and run the whole thing backwards. The copying we are talking about here is the bitwise copying where 0 is copied to 0 and 1 to 1 - we do not actually clone a quantum state, since this is impossible. And this is where the 'V' shape returns: that's why the addition has that shape - to get rid of the garbage at every stage of the addition.

The process can be seen in Figure 12.4, with five registers at the input. It computes on the first two, then on the second two, and continues like that all the way to the end, where we have the result. Now we just copy our result, then we run exactly the same network in reverse, so we undo what we computed. At the end we have the auxiliary reset to zero. Just as another reminder: this is a standard trick from reversible computation. The only difference with quantum computation is that the input can be in a superposition of as many x's as you like.

Now to a brief example where quantum computing is more efficient than classical computing. There are many examples, but I have chosen one that shows relatively quickly that quantum computing is more efficient. It is Deutsch's problem: this is a very nice problem where one can see how superposition works directly to give an advantage in quantum computing. The problem is set out as follows:

Let f : {0, 1} → {0, 1}

There are two possibilities:

a) f is constant
f(0) = f(1)

b) f is varying
f(0) ≠ f(1)

Task: by computing f only once, determine its type.

The function, f, is a binary function. That means that the variable is also binary, which means we can have f (0) or f (1) and the value of these can be either 0 or 1. So we have four possibilities. But let's suppose we don't actually care about the four possibilities: suppose we want to know only one thing in this case. We have two possibilities: either f is constant, which means the value of the function at zero is equal to the value of the function at 1, or we have f is varying; which means that f(0) is not equal to f (1). The problem Deutsch proposed is to compute f only once and determine its type. It is simple to see that you cannot do this really with a classical network: we have an input x and some box that performs computation f and an output f(x). First we have to put zero in to get f(0) and then we have to put in 1 to get f(1); we definitely have to do it twice.

But now, in Figure 12.5, we can see the magic of quantum mechanics. In the figure, time runs downwards.

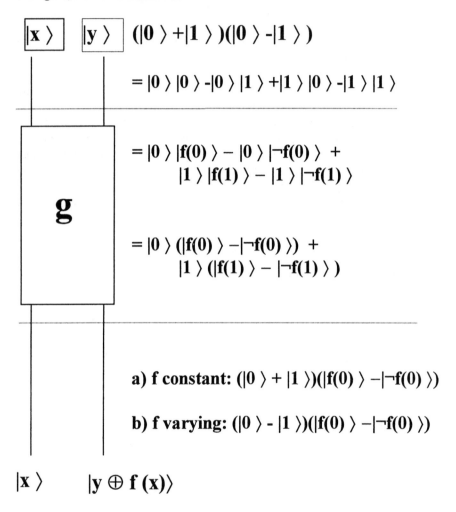

Figure 12.5: the quantum mechanical path for solving Deutsch's Problem.

We have two inputs x and y, and we have a gate g. We keep the first qubit unchanged and in the second qubit we compute the function y + f(x).

We only need to know the basic quantum mechanics to understand how this works: the input is a superposition, and there is a difference in phase between the first qubit and the second qubit. The second line expands the superposition:

notice the four phases: the second and fourth element have minus signs. Now we compute the gate g.

Once we have the result we can factor this. By pulling out |0> from the first two terms what stays inside is f(0) - Not f(0). By pulling out |1> from the last two terms we have f(1) minus not f(1) remaining.

Now comes the crucial moment: if f is constant, then the first qubit is in the state (|0> + |1>) - and it doesn't matter what state the last qubit is in. On the other hand, if the function is varying, the first qubit is in the state (|0>-|1>). So these are two orthogonal states; we have distinguished them with 100% efficiency - at least theoretically.

To conclude, I shall reiterate that all the elementary arithmetic operations are possible, performed in a reversible way borrowed from reversible computation theory. I also showed the trick of not developing a load of garbage, by introducing this 'V' shape into the network. Then we showed why quantum computers are faster. Of course, with only two qubits we can't show an exponential increase in speed. And remember: all the considerations that have been discussed are entirely free of errors - in order to make this work in practice, we have to take decoherence into account.

Chapter 13
Quantum Cryptography

Richard J Hughes
Physics Division, Los Alamos National Laboratory

At Los Alamos we have rather a different perspective - in some ways - to those working in a strictly university environment. We have to do things that are more application oriented in the long run, and quantum cryptography is an application that, at least from the point of view of physics, is feasible today.

Cryptography is essentially about three people. Alice wants to send a message to Bob without allowing Eve to eavesdrop on the communication (Figure 13.1). In conventional 'symmetric' systems, Alice and Bob share a key: a sequence of random bits that enables Alice to encrypt the message. They face a standard problem: if Alice can't send a message to Bob without it being intercepted, there is the problem of how to get the key across in the first place. It is this problem of key distribution that quantum cryptography deals with. Quantum cryptography has very little to do with encryption algorithms - it's to do with key distribution by single photon transmission. The potentially attractive features are its immunity to eavesdropping (or you can at least detect if it's going on), and the fact that it might make some things easier in the cryptographic world.

To be concrete, I'll begin with an example of some cryptographic key material that we produced for Alice and Bob through single photon transmission in our 48 km optical fiber experiment. This encryption procedure (known as the 'one time pad') is a conventional procedure, nothing to do with quantum mechanics.

The quantum mechanical part is the Quantum Key Distribution. This requires that Alice and Bob can generate truly random numbers, perform quantum comparisons, correct errors and squeeze out any information that Eve might have collected (Figure 13.2).

Alice prepares a secret random bit sequence (by, for example, digitizing electrical noise) and transmits it to Bob in the quantum states of photons over a quantum channel. Bob measures it, and sends back information over a public channel about which photons arrived. This allows Alice and Bob to get rid of

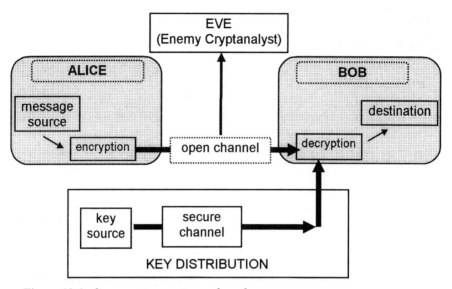

Figure 13.1: the quantum cryptography scheme.

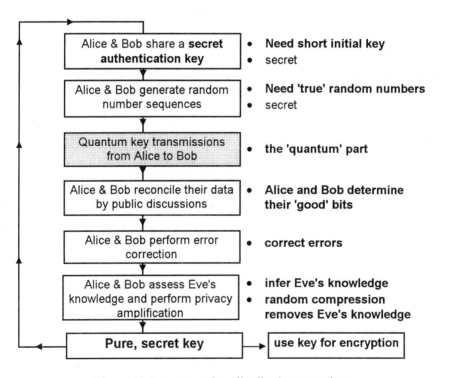

Figure 13.2: quantum key distribution procedure

errors and distil out a final random subset of numbers. Its security is ensured because Alice is sending the photons one at a time: they can't be split or cloned and neither can they be measured without modifying some of their properties, thanks to the nature of the quantum states. There are, of course, more sophisticated attacks that are possible. Eve may sit on both channels and pretend to Alice she is Bob, and vice-versa. This can be overcome by ensuring that the messages in the public channel also require a short authentication key. Other attacks, known as 'coherent' attacks, may be more difficult to overcome; this is indeed an area of current research.

In the end, Alice and Bob share an error-free key, some of which is used to replenish the authentication key and the rest is used for encrypting messages. So, in the example of Figure 13.3, a phrase is encrypted by performing an Exclusive-OR operation on each ASCII character with 8 bits of the quantum key material. This sum is then transmitted in the clear, and Bob decrypts the string using the same operation with the same key.

The quantum channel used for transmission of the key bits can be an optical fiber or the atmosphere. Using wavelengths around 1.3 microns and 1.55 microns in an optical fiber is possible, but there are practical considerations: it is not possible to simply buy photon detectors for these wavelengths, although various research groups have shown that avalanche photodiodes can be used to count photons at these wavelengths.

We have performed quantum key distribution over a network of 'real world', underground optical fibers of over 48km for our experiments. Sceptics who have said that interconnections would ruin the beautiful laboratory-based experiments have been proved wrong: our experiments have worked over this distance, despite the interconnections. That would mean, for example, that central Washington could be connected to most of the outlying government agency buildings over a secure quantum cryptography network.

Our latest experiments have shown that the atmosphere is a practical quantum channel. There is a ground to space infra-red transmission window at 770nm. This is a convenient area because one can buy single photon detectors, based on silicon technology, for this wavelength.

Firing photons down a neutron time-of-flight channel in an experimental area of the Los Alamos National Laboratory, we have succeeded in carrying out a 1km free space experiment. The photons were sent out to a mirror 0.5km down the channel, and reflected back. The experiment worked well at night. (More recently, we have demonstrated a point-to-point quantum key transmission over the 0.5-km path in daylight.) We used the experiment to encrypt an image of St Louis airport (Figure. 13.4). Combining each pixel with a byte of keyword in an Exclusive-OR operation encrypts it, Bob is then able to use the same key to decrypt it.

We hope to extend this successful operation to 2km and eventually to 7km; assuming that works we would like to do a space experiment. The idea of receiving one photon that has been sent up from the ground to a spacecraft seems almost impossible, but there are three things that will work in our favor. You

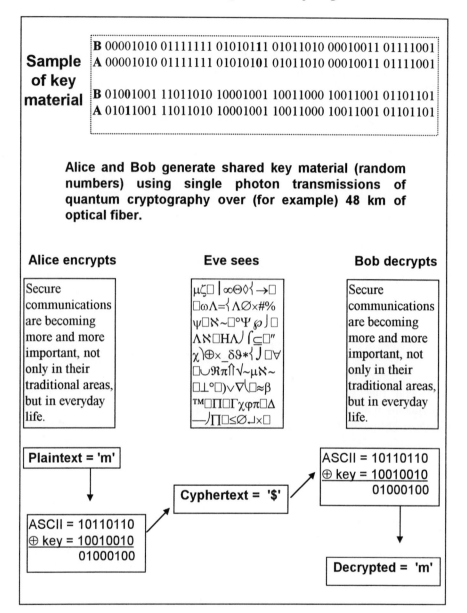

Figure 13.3: an example of Quantum Cryptography.

know when it is going to arrive (you can send out a bright timing pulse); you know the wavelength it has, and so you can use a narrow wavelength filter; and you know the trajectory, so you can use a small-angle receiver.

Quantum Cryptography

Alice encrypts by adding a word of her key:

00111101 to each pixel

Encrypted Image

Bob decrypts by subtracting a word of his key:

00111101 from each pixel

Figure 13.4: encryption and decryption of an image using free-space quantum cryptography (no error correction).

Even in very bad conditions we expect to be able to transmit at 10MHz (using one photon per pulse), and produce a key at 0.5 kHz 300 km up in space. Since the typical satellite would be in view for approximately 8 minutes we can assume 1 minute of QKD transmission time: enough for several days worth of session key generation. This could, then, be a useful application.

There are still a number of improvements that could be made to quantum cryptography systems. High on the priority list is a proof of absolute security, and an understanding of the relationship between the Bit Error Rate and Eve's knowledge. Ideally that would be expressed by an understanding that "a BER of x% implies Eve's knowledge is less than y bits".

For efficiency, we need practical, integrated reconciliation, error correction and privacy amplification protocols. We need to minimise the necessary initial secret key bits and the public channel bandwidth, and maximise the output of key material.

In terms of technology we need true single photon sources (possibilities are down-conversion, the 'photon turnstile' and the 'photon gun'), integrated optics for the receiver and transmitter, and a robust low noise and high efficiency single photon detector (especially in the infra-red region of 1.3 and 1.55 micron wavelength). We also need physical random number sources, which are of the utmost importance in this field.

In summary, I have shown that quantum key distribution is indeed feasible for optical fiber communications of less than 100 km and for line of sight optical communications through free space. These developments are in the prototype stage.

In order for quantum cryptography to be used, we need to convince cryptographers and potential end-users that QKD is secure. It must also do something 'better' than conventional key distribution, for example, demonstrate ease of use coupled with high security. It may then be adopted for use where couriers are impractical and the conditional security of public key methods is inadequate. It may well also be used in satellite communications.

Chronology of Quantum Cryptography

PREHISTORY
Weisner ~ 1970
- quantum information storage
- 'quantum money'

↓

FIRST QKD PROTOCOL
Bennett & Brassard (1984)
- quantum information transmission
- key generation

↓

NEW QKD PROTOCOLS
Ekert (1991); Bennett (1992)
- security based on Bell's inequalities
- any pair of non-orthogonal states

↓

FIRST QKD EXPERIMENT
Bennett et al (1992)
- proof-of-principle
- 30 cm

↓

QKD FIBER EXPERIMENTS
Rarity,...Townsend et al (1993 - present)
Franson et al (APL)
Gisin et al (Geneva)
Hughes et al (LANL)
Bethune et al (IBM)
- interference (polarization) schemes
- 10km...48km
- multi-user polarization stability
- fiber networks

↓

FREE SPACE EXPERIMENTS
Franson et al (1996)
Hughes et al (LANL)
- polarization schemes
- 1km range achieved
- surface to satellite?

↓

IN-DEPTH SECURITY STUDIES
Mayers & Yao
Lo
Mor
Brassard et al
Lutkenhaus
- security against arbitrary quantum attacks

Section III

A Perspective for the Future

The following chapters attempt to outline the state of QCC research worldwide, and examine some of the implications for the future of this field.

Chapter 14
Realizing the Potential of Quantum Information Processing

Michael Brooks

14.1 Prospects for Quantum Computing

The quantum computing system heralds the first fundamental change in the nature of computing since the age of modern computing dawned with the Turing machine. The natural parallelism of superposition enables, in principle, computations to be carried out that are not now, or ever will be, practical with classical systems. For example, the application to factoring large numbers would have a revolutionary impact on cryptography. But in practice other applications, many as yet unrecognized, are likely to have an even bigger impact; for example, the ability to extract a single entry, fast, from a large database of unordered information would have a major impact on fields like feature (e.g. fingerprint) recognition, language translation and speech recognition by corpora matching, and the many-body problem (e.g. air traffic control) to name but a few. However, industrial economics dictates that a new approach is unlikely to obtain a market hold unless and until the benefit is very significant. In early applications the quantum computer might be used as a type of co-processor to the established classical hardware. But these will not take off in commercial terms until the number of qubits that can be processed provides a system that has indisputable benefits over the classical approach. This implies that there would be no mass market for normal computational purposes until the systems can handle large decimal numbers. (The actual number of qubits needed will probably be determined as much by the error correction requirements as by the numerical size of the task). Machines with registers for tens, hundreds, or even thousands of qubits will be required. It is likely to take many years, say 10–20 years, to reach that development stage.

While it is now beyond doubt that a quantum computer capable of handling a few qubits can be built (and has been demonstrated in a primitive form using the NMR technique), it is by no means certain that such a system could ever be built in practice (the 'house of cards' effect); and, if it can be built, it might require the use of cryogenic techniques to reduce noise and decoherence, with all that that implies in terms of system cost. So one has to conclude that while the ultimate goal of a quantum computer handling large numbers is one with profound implications for ultimate industrial application, there is no prospect of a general purpose quantum computer reaching the market for many years to come.

14.2 Prospects for Special Applications

There are certain applications of quantum information processing, perhaps even of quantum computing, where a system with a limited number of qubits, or simply a system with reduced quantum uncertainty, would have valuable applications. A few of the most important applications for industry are mentioned below:

14.2.1 Quantum Simulation

The importance of simulating chemistry at the individual particle or molecule level is well recognized (the subject of the 1998 Nobel Prize in chemistry for which both the two recipients originally came from Europe). For example, the subject is of much interest to the pharmaceutical industry. But the complexity of the calculations limits the size of the molecules that can be simulated, and there is strong market pressure for systems that can simulate more complex systems. This might create a specialist industrial market pull for quantum computing systems that can handle, say, 10 or more qubits. And it must be recognized that there will be a ready, if limited, market for such systems in many, perhaps most, academic physics and chemistry laboratories throughout the world.

14.2.2 Limited Qubit and Reduced-Noise High Precision Applications

As with any new scientific development there are likely to be valuable applications that are not at first identified. As well as the use of quantum computing applications, potential applications have been identified which involve limited numbers of qubits or high-precision sub-shot-noise applications and so are likely to become of industrial exploitation significance within the short term. For example, in metrology, there exists the possibility for improved precision atomic clocks and highly accurate measurement of distance for precision navigation. Quantum gyroscopes, where very large increases in sensitivity are predicted, have implications for the detection of minute gravitational deviations on Earth from Space, as well as for scientific missions.

14.2.3 Secure Communications

In the short term, the use of quantum information processing for secure communications would seem to be the largest and most promising potential market. The weak point in the current common method of using codes that depend upon a key that is changed frequently, lies in the transmission of the key from the generator of the message to the recipient. Quantum communications provides a safe method of transferring a truly random key, depending essentially on the transmission of a string of one half of a pair of entangled photons using some suitably random property (say polarization) to provide the binary key, the other half providing the same key to the generator of the message. Because an eavesdropper will destroy the entanglement, it is possible to elaborate the process using the random nature of quantum properties, not only to send a key in absolute security but also to apply statistical tests, which will demonstrate if an attempt to eavesdrop has been made. The method has been demonstrated to work and is essentially ready for industrialization, in the first place for use over distances such as that of a city. Bodies like financial institutions, as well as security and defense agencies, are becoming interested in the deployment of such systems. The potential market is obviously immense. There can be no doubt that this market will give a boost to the whole subject and provide the usual stimulus to improvement provided by real user experience.

14.3 Meeting the Needs of the QIP Field

It is self-evident that the subject of QIP is of very considerable potential importance for many aspects of the computing and communications field. This is perhaps the most fundamental change that has appeared in the basic underlying technology of the field since the digital computer age was introduced some 60 years ago. Whether the very long term goal of a quantum computer, such as one that can overcome the barrier for classical computers handling large numbers on certain types of problems, can be achieved must remain to be determined as R&D work progresses. But there is already such promise for valuable applications along the road, such as that for secure communications, that it would be dangerous to leave the field undifferentiated from other emerging challenges. It has the rare property of being seen as exciting both by the pure scientists engaged in the field and by those who recognize the value of the potential applications. Moreover the science involved is so close to fundamentals that it is highly unlikely, to say the least, that new science as well as ideas and applications, as yet unthought of, will not emerge. Occasionally an opportunity occurs in the early days of a new technology when a relatively modest investment of funds can have a disproportionately large effect on its development, setting off a dynamic and self-supporting expansion leading to the establishment of what is essentially a new industry. There seems to be evidence to show that the opportunity exists today to stimulate the growth of a critical mass of knowledge, effort and expertise and create a new industry in these emerging technologies.

The very rapid growth of interest in the subject, involving various disciplines, can be judged by the large number of recent articles, not only in the scientific and technical press but in the general news media as well. The work of the theoretical quantum physicists has been opened up by the experimental work carried out over the last ten years or so. Old arguments about fundamental principles have been subjected to the test of experimentation, and as the physics community involved has grown, so have the novel practical solutions propounded to overcome what were previously seen as fundamental obstacles. Skeptics have joined the field only to find themselves opening up new lines leading to the solution of the very problems they foresaw. Perhaps the most important developments of recent years have been the entry to the field of people from other backgrounds; computer scientists with ideas about error correction and algorithms for quantum processors; chemists for work with the NMR equipment approach; and even mathematicians are finding the subject so interesting that they are taking a renewed interest in it. It is this convergence of the cultures that promises a rapid acceleration of the field if the disciplines can be enabled to interact.

14.3.1 The Need for Academic Focus

While the numbers have grown rapidly, and the number of different disciplines attracted to the field has increased, the field remains thinly spread in many places. In Europe, with the exception of the three or four main centers of excellence, the workers are spread thinly in small groups with quite uneven distribution, needing the stimulus of interacting with their colleagues both in their own discipline and in the others that are contributing to the field. The field is unusual in the relative lack of understanding of the subject in disciplines outside physics. The apparent 'weirdness' of quantum physics creates barriers to understanding that have to be overcome if those schooled in classical ideas are to contribute fully to the development of the systems. The need to bring the different disciplines together has become very apparent in recent developments stemming from a variety of different disciplines. Without inhibiting the individual approach that must remain important in such an immature subject, the time has come when some degree of structure to the research effort can help both to enable the disciplines to interact, and to provide support of interaction for the far-flung community.

14.3.2 The Need for Industrial Focus

The active industrial interest remains very sparse, but this may be expected to change dramatically once it is known that live commercial experiments are taking place, as can be expected to happen in the next few years, for example with the financial community trying out the secure communications systems. This is a field both with very long-term aims and expectations for major exploitation, and with the potential for valuable exploitation along the path. It is desirable to create some forum where industry can be given the opportunity to interact with the research workers to understand what the new technology requires, what the problems and pitfalls are, but also what the coming opportunities are. The chemical and pharmaceutical industries may be expected to want to experiment with quantum simulation well before the technology has reached the point of providing them with valuable results. And electronics and avionics systems firms concerned with precision measurement and navigation should be aware of the developing technology. Moreover, though industry may not wish to take an active role in many of the experiments where application is a long way off, their presence in regular forum meetings will ensure the experimental work takes advantage of the preferred manufacturing options when they arise. For example, it would be desirable to have the semiconductor industry playing a watching brief in such a forum. And it would be desirable for both sides if the microsystems structure expertise could be brought to bear on replicating laboratory pre-prototypes. Such a forum might help to ensure that European based industry is at the forefront in the application of the work, as new application opportunities emerge. This is a field where new scientific and application ideas are appearing all the time, and it is desirable to help industry to pick up the awareness of these changes from the scientists themselves without too much delay. And since it may be expected that new firms will appear to take some of the technology into applications, it is never too early to expose the venture capital world to the emerging opportunities through special industrial forum meetings.

14.3.3 The Need for Awareness

The opportunity to enthuse the young, both school children and undergraduates, with the excitement and opportunities in this field should not be missed. This argues for a degree of central approach to orchestrate such a campaign and to encourage and help all countries to take advantage of material stemming from others. There is a particular problem where the young are turning away from subjects like physics and information technology. This is reflected in the fall in university entries in these subjects in some countries. The QIP field is highly suitable to act as an example to revive interest in the young at the intellectual and exploitation challenges that are to be found by working at the frontiers of physical science.

Chapter 15
The Role of Europe

Michael Brooks

15.1 The Pioneering Stage

The subject of QIP may be said to have had its roots in Europe in the work of the great European physicists during the first 20 years of this century. Then when the subject became of interest again in the 1970s and early 1980s US work became of importance, notably that of Charles Bennett and Richard Feynman. But Europe was always well represented both in the fundamental thinking and in fundamental experiments. Indeed, in experiments Europe was leading in interferometry with individual quanta (electrons, neutrons, photons). These techniques are now important in QIP. In 1985, David Deutsch in Oxford moved the subject forward, and thereafter the theoretical advances have developed on both sides of the Atlantic. Quantum gate development has proceeded on both sides of the Atlantic. The USA has led in the work on algorithms (notably at Bell Labs), but the crucial work on Error Correction has been shared between work in Europe and USA. Experimental work has developed in the 1990s in parallel in USA and Europe, with significant teams in the USA, for example at Los Alamos and MIT, but with centres of experimental and theoretical work developing in Europe at, for example, Geneva, Innsbruck, Oxford and Paris. The strong optical physics community in Europe has helped to form a strong foundation for both experimental and theoretical work. The ion-trap quantum computer idea was first proposed in Innsbruck. NMR techniques, with the practical demonstration of quantum algorithms, seem to be being pioneered primarily by the two centres in MIT and Oxford.

15.2 Today

Europe can be seen to be well up-with-the-hunt in both theoretical and experimental work. It is at least as likely that advances will be made in Europe as in the USA. The one area where the quoted advances seem to stem largely from the USA is in algorithm work, but the strength of the fundamental computing science logic community in Europe suggests that this may not remain the situation for long. In applications developments, experimental quantum cryptography work was pioneered in Europe. The work at Innsbruck, Rome and at Geneva Universities on teleportation and secure long distance (over many kilometres) communications has drawn the attention of the world. Recent work at Innsbruck on three quantum particle entanglement is another first for Europe.

15.2.1 Multi-Disciplinary or Trans-Disciplinary Nature of the Community

At this time, the subject is developing at an ever-escalating rate, and new people are entering the field, not just through the route of post-graduates and post-docs, but through conversion from other work. The field was, until recently, largely populated by quantum physicists. But now computer scientists, mathematicians with experience in fields like probabilistic computing and chaos theory, chemists with NMR experience, and semiconductor physicists and engineers are being drawn to the subject. The interest in topics that are closely related to the field, like the fundamental nature of information, is growing. And on the edge, though stimulated by some of the thinking of people like Roger Penrose, there are some 'cognitive scientists' who believe they find a model for the operation of the brain in quantum computing. The field is becoming strongly multi-disciplinary.

15.2.2 The Numbers Involved

It is difficult to judge the total community, if only because some of them are also engaged in other different work. But there is no doubt that at least several hundred graduate scientists, mathematicians and engineers are engaged in the field in at least 20 countries of Europe. It would appear that the numbers are growing fast and may exceed 500 in Europe within a few years. It is not clear how this compares with the numbers in the USA, but the support of defence and security agencies there (ARO, NSA, NASA, and DARPA) provides considerable support for the work. Moreover, it is known that NSF are now considering launching a programme of work. In total the support from these agencies probably amounts to some $15 M per annum, and growing. So the total number of workers in the USA probably exceeds the European figure but, at the moment, perhaps by not a very large factor. The existence of the DARPA and NSF programmes will certainly act to encourage more centres and individual workers to be drawn in.

Elsewhere, there are teams in Japan but it would not appear that the scale of the work compares with that in the USA and Europe.

15.2.3 The Geographical Spread

Most of the countries of Europe have at least one or a few teams working in the field, no doubt encouraged by the cryptographic and secure communication interests of all countries. The situation can be described as "lumpy". There are strong teams in certain centres, for example in Austria, France, Italy, Switzerland and the UK, with work known to be going on in countries like Belgium, Denmark, Germany, Sweden and Finland, and no doubt elsewhere in Europe. The picture in Germany is changing fast with a new programme recently launched by the DFG. A call for proposals has resulted in some 60 applications, and it is expected that at least 20 projects will be launched in 1999. And it is worth noting that Israel has several centres producing excellent work. There is also work going on in Eastern Europe, notably Poland and Slovakia. Groups in the Czech Republic, Croatia, Hungary, and Slovenia have expressed an interest in joining the emerging field of QIP.

15.2.4 The Industrial Scene

In the USA some of the pioneering work has stemmed from the long-range research laboratories of the large ICT firms, such as IBM, and Bell Labs of Lucent (Technologies). In Europe the large telecomms firms (France Telecom, BT, Swiss Com) take an active interest but the pioneering work has largely stemmed from the academic laboratories. These large firms have a double interest, stemming from their work in photon optics and their interests in secure communications. It may be assumed that all the telecomms firms take a monitoring interest in the work, and probably most support some academic work. (In the USA the security and defence agencies probably fund much of the work outside the industrial laboratories, and some of the academic work in Europe is supported by the equivalent bodies here). Microsoft has a small team in Seattle, but are not known to support any work in Europe. Hewlett-Packard support relevant work in their long range research laboratories in Europe. In Europe, the ICT division of Elsag-Bailey SpA, based in Genoa, has supported in-house research, and academic work for a number of years. Their support for seminar activities for the outstanding physicists in the field from around the world for some years has made a major contribution to co-operation and the development of the subject. The Star Lab of Riverland nv is notable as a relatively small firm which has established a small research team in Brussels and also supports work elsewhere in Europe and the USA. There is known to be a small Polish company which takes an interest in the subject.

15.2.5 Summary of the European Scene

The scene in Europe is one of rapidly growing interests from several disciplines, and a number of strong academic centers. But the current community is spread very thinly and unevenly across Europe. With a few exceptions, the industrial scene is sparse outside the main telecomm firm laboratories, though it may be expected that all the telecom and electronics firms with interests in secure communications are, at the very least, monitoring the scene carefully.

Chapter 16
Quantum Computing and Communications: A View from the USA

Colin P. Williams
Jet Propulsion Laboratory, California Institute of Technology

16.1 Introduction

In addition to identifying what I consider to be some worthwhile technical directions for the field, I thought it would be valuable to talk a little bit about my impressions of the quantum computing and quantum communications programme underway in the United States. In particular, I would like to focus on what works well and what does not and to see if there are any gaps that could be filled by a new, coordinated, European effort.

Although this is not widely known, the American NSA-ARO-DARPA research programme in quantum computing was kindled, indirectly, by European speakers and communicated to the U.S. funding agencies by Jon Dowling who was, at the time, a physicist in the U.S. Army Aviation and Missile Command, and who has recently joined my group at JPL. Jon had attended the 1994 IQEC and ICAP meetings, where British Telecom revealed their transmission of a quantum cryptographic key over 10km of optical fiber, and Artur Ekert announced Shor's discovery of the factoring algorithm, respectively. Upon making these breakthroughs known to the Army Research Office, it was decided to have an ARO workshop on quantum cryptography and quantum computing in Tucson in the winter of 1995. Many of the key researchers in quantum computing were at this workshop, as well as Keith Miller and others from the National Security Agency (NSA). This workshop was the genesis of the ARO-NSA collaboration on the promotion of quantum computing research. Working closely with ARO and the NSA is the Defense Advanced Projects Research Agency (DARPA), which funds a collaborative effort between M.I.T.-Caltech-USC called the QUIC (Quantum Information and Computation) programme (a part of the DARPA ultrascale computing programme). There are now plans to have an NSF centre in quantum computing based at M.I.T.

16.2 What Works Well

The NSA, ARO and DARPA agencies are all highly mission-oriented. I believe that this has helped focus the research programme in quantum computing without compromising innovation, or opportunistic exploration of new ideas and phenomena. To achieve this, it has been a critical factor that Keith Miller (NSA), Henry Everitt (ARO) and Sunny Maynard (DARPA) have maintained a very frank dialog with the research community in the face of rapidly changing, and unanticipated, priorities and possibilities. They seem to have struck a very good balance between fundamental physics research and applications-oriented research.

One thing that I really like about Henry Everitt's approach to managing the ARO programme is that his periodic meetings end each day with a panel session at which the days speakers field (usually pointed) questions from Henry and the audience on the relative merits of the ideas advanced. I think these public debates have been immensely helpful in clarifying the state of the field to everyone concerned. These discussions also expose the goals of the funding agencies so that researchers can submit more responsive proposals.

16.3 What Does Not Work So Well

I think one of the inefficiencies of the American approach has been the lack of encouragement of true collaborations between theoretical computer scientists and physicists. These are still very much two communities who meet periodically to exchange a volley of academic papers but then retire home to pick up their work again. As a result, the styles in which the quantum computing papers of computer scientists and physicists are written often do not foster mutual comprehensibility. Indeed, in his paper proving the optimality of Grover's algorithm, Christoph Zalka, a physicist at Los Alamos National Laboratory, ends with the sentence "I apologize to mathematicians and computer scientists who have suffered through this paper, looking in vain for theorems, lemmas, corollaries, ... Probably it is of little comfort to them that it took me weeks to understand their presentations on quantum searching" [1]. I think a new European effort should confront this issue head-on. In part, it is a relic of the education process: physicists and computer scientists tend to specialize indifferent branches of mathematics, but it is also sociological: the community of physicists tend to emphasize simple approximate models that are analytically tractable whereas the community of theoretical computer scientists tend to emphasize formal proofs and rigour.

In America I am disappointed that the quantum computing field is composed of only physicists and theoretical computer scientists. There is little representation from researchers in operations research, artificial intelligence, and computer systems engineers. I believe that participation of these other types of expertise will be vital to grow the quantum technologies from fledgling prototypes to full-scale devices and to stimulate a research focus on practically significant computational problems. Moreover, it is a two-way street. For example, for several decades now artificial intelligence researchers have been developing 'machine discovery'. Apart from some true discoveries in pure mathematics and chemical synthesis pathways, these efforts have not discovered anything that was not already known. Quantum computing is a perfect test-bed for machine discovery: it is a fairly well circumscribed, mathematics-based field, and is therefore fairly easy to formalize. I have started to pursue this line of thinking in my own work on quantum circuit design using genetic programming [2], but I would welcome more A.I. people entering the quantum computing field.

16.4 NASA/JPL

What about NASA and JPL then? In 1997, I joined the Jet Propulsion Laboratory (JPL) to help start a NASA effort in quantum computing research. NASA is also a mission-oriented agency centred around four 'Enterprises': Aeronautics, Earth Sciences, Human Exploration and Development of Space, and Space Sciences. JPL's role is primarily deep space exploration. In addition to these avowed mission enterprises there are other, more informal, objectives too. Much of NASA's research work eventually stimulates growth in American industry. The more publicized space missions inspire the people, and encourage interest in science amongst school children. It is hard to put a financial value on NASA's activities but there is clearly a large social impact as evidenced, for example, by the Mars Pathfinder website being the most visited website in history.

Our interest in quantum computing at NASA/JPL stems from the need to solve hard, supposedly 'intractable', computational problems on a routine basis. Detailed mission planning, deep space network communications scheduling, and spacecraft design optimization are just a few examples of the kinds of NP-complete problems that we face. So far, there is no known classical computing technique that can solve such problems using polynomially bounded resources in the worst case. Today we employ a small army of computer scientists to work on developing the very best classical algorithms for performing these computations. As we move out into the next century, NASA envisions a huge increase in the number of concurrent space missions. In such a climate the computational demands we will face will become significant factors impeding progress. Moreover, as our spacecraft become more remote from Earth, the speed of light limited communication delays force us to place more autonomy on-board. To function autonomously, such as to re-plan a series of observations, and data analyses in real-time during a fly-by of an astronomical body, we need to bring massive computational power to bear under extreme constraints on allowed mass, time and power. Like the ARO, and DARPA, we would like to know if quantum computers can provide any help.

Historically, NASA has been one of the key innovators of computer technology in the United States. To give just a few examples, JPL has designed and built massively parallel computers, hypercubes, and neural network hardware and we currently invest in superconducting electronics based computers, quantum dots, nanotechnology, and reconfigurable, fault-tolerant hardware. These research activities rarely make the headlines as the space missions do, but computer technology research plays a critical supporting role for our primary mission objectives.

We recognize, however, that there is more potential from quantum technology than just computation. Moreover, certain non-computational devices appear to be more feasible to implement within the near future. In an effort to elucidate new ideas, and to push the envelope on existing ones, in 1998 JPL organized the first NASA International Conference on Quantum Computing and Quantum Communications, held in Palm Springs, California. We imposed an a priori structure of the meeting that was designed to solicit feedback on how quantum computing might impact NASA mission objectives in computation and communications. I am delighted to say that NASA really obtained great value for money from this conference. In particular, four key ideas emerged that caught NASA's attention: quantum gyroscopes, improved precision of atomic clocks, the first steps in quantum algorithms for tackling NP-complete (structured) problems and the potential for Earth-to-space quantum key distribution (QKD). The proceedings are published as Volume 1509, of Springer-Verlag's Lecture Notes in Computer Science.

The quantum gyroscope idea was discussed by Jon Dowling [3]. If two particles, entangled in a special way, are fed into a two-port Mach-Zehnder interferometer, the phase sensitivity scales as the Heisenberg limited $\partial\phi = O(1/N)$ where N is the number of particles incident per unit time. In a one-port device the phase sensitivity scales, at best, as $\partial\phi = O(1/\sqrt{N}]$). Calculations suggest that the two-input port optical quantum gyroscope ought to be about 10^8 (one hundred million) times more sensitive to rotations than a one-input port optical gyroscope. Such an increase in sensitivity will be vital in future missions that must navigate dense asteroid fields or debris thrown off from comets. Moreover, quantum gyroscopes might also enable new classes of Earth-observing missions based on detecting minute gravitational deviations, and may help in pointing the arms of massive space-borne interferometers that are planned to obtain high resolution images of planets orbiting stars in other solar systems. Quantum gyroscopes will also facilitate new tests of general relativity. Thus, for NASA, the confluence of quantum computing techniques and optical gyroscopy has the potential to be 'mission-enabling', i.e., to allow us to conduct missions that are impossible using classical techniques alone.

Another highlight of the Palm Springs meeting, for me, was Lov Grover's presentation on structured quantum search [4]. Subsequent discussions led to the collaboration between Lov Grover, Nicolas Cerf (then Caltech, now

Brussels) and myself that culminated in our 'Nested Quantum Search' technique [5].

Similarly, Daniel Abrams (MIT) presented a provocative paper showing that if quantum mechanics is nonlinear, then quantum computers can solve NP-complete problems in polynomial time [6]. Of course, it seems pretty unlikely that quantum mechanics IS nonlinear, but the question has not been decided definitively experimentally so far, and so the result is indeed intriguing and just the sort of stimulating new idea for which we had hoped. Dan now works part-time at JPL on new quantum algorithm development. No less provocative was Richard Jozsa's idea of quantum computing by leaving the computer turned off (basically a re-interpretation of the quantum eraser experiments) [7].

16.5 Lessons Learned from Experience with NASA

Let me summarize some principles that I have learned from my experiences at getting a quantum computing programme going at NASA/JPL.

I believe it is important to identify short term, medium term and long term goals. For a quantum computing and quantum communications programme to succeed, it may have to span several decades. To convince funding agencies to maintain their support they need to see some tangible progress along the way. At JPL although my personal ambitions lie in quantum algorithms per se, I see the quantum gyroscope and Earth-to space quantum key distribution, as serving as valuable stepping stones along the way.

I found the Palm Springs meeting to serve a greater purpose than merely reporting on recent advances in the field. By imposing an a priori structure on the conference that challenged people to address to our mission needs, enough of the researchers contributed responsive papers that we obtained real value from hosting the meeting. Moreover, by not insisting on complete papers too far in advance, a high proportion of novel ideas were presented.

16.6 Opportunities for Europe

Let me turn now to the question of the quantum computing in Europe. A good question to ask would be 'What is unique about doing quantum computing research in Europe?'. Are there information processing goals that only Europeans care about? For example, machine translation based on word-substitution is a trivial computational problem, but true semantic-based machine translation is extraordinarily difficult. Perhaps quantum computers for machine translation is a worthy goal? Conceivably, it could lead to a technology of translating telephones and translating photocopiers etc.

Secondly, I would ask 'Are there specialists within Europe who could bring new ideas and technical challenges to the field?'. One thing that comes to mind is the relative strength of the pharmaceutical industry in Europe. Could Europe find 'grand challenge' computational problems, such as drug design, that would encourage a focussed quantum algorithm development?

Certainly, I would like to see Europe develop its own vision for quantum computing and not merely replicate the priorities and investment strategies used in America. As I said in Helsinki, my impression is that the bulk of the funding in quantum computing in the U.S.A. is geared towards hardware efforts. For NSA this is completely understandable: they already have a quantum algorithm (Shor's factoring algorithm) that they really want to implement. However, for the rest of us, I think it is far from obvious that we know that there are enough interesting things to do with quantum computers to warrant a massive investment. Fortunately, computer scientists have fairly low overheads in comparison with experimental physicists. So a little funding in computer science research goes a long way. I would really like to see a strong European effort in quantum algorithms research that was focussed around a few key grand challenge computational objectives. In short, we need to find more 'killer' applications.

16.7 Commercialization

I believe that quantum cryptography could be commercialized at this point. As Richard Hughes (LANL) has pointed out, and as I have said before too, the QKD technology is sufficiently well prototyped that it is feasible to create an intra-city quantum network. A more ambitious European goal would be the challenge of building a trans-European quantum communications network.

With regard to Stan Williams' excellent - albeit sobering - account of the corporate viewpoint [see chapter 9], I would like to add a counter-balancing comment. First, we do not yet know whether the computing needs of 2050 and beyond will be dominated by the kinds of computational tasks at which quantum computers will excel. So predictions of potential markets and market share become more questionable the more distant the predictions. Turn the clock back 50 years and try to predict the path of technological development that we actually followed from 1948 to 1998. In 1943, Thomas Watson, then chairman or IBM, said, 'I think there is a world market for maybe five computers.' Likewise, Popular Mechanics ran a bold article in 1949 that envisioned the dramatic advances that were expected in computer design. Their conclusion was that 'Computers in the future may weigh no more than 1.5 tons'. Whilst this statement is of course correct, it falls a tad short of capturing the true scale of the computer revolution! Can we blame these myopic visionaries? Who could have foreseen the technological basis of, and social need for, personal computers, word processors, spread sheets, games, image processing, multi-media presentation tools (when they work) etc. If history has taught us anything it is that we are pretty bad at anticipating future technology, and the needs of society. Moreover, there is always the potential for spin-off industries along the way. In a sense, we might already be witnessing this with quantum cryptography: QKD was inspired by thinking about the peculiar features of quantum information and now it might become a stand alone industry in its own right.

Today it seems to be very premature to be thinking about a quantum computing company. What would the products be? If the business basis of the company is the acquisition of core patents, then that strategy is only as viable as your desire to prosecute patent infringement. However, I agree that the situation could change rapidly if a scalable quantum computer architecture is demonstrated, and if computer scientists find a wider range of 'killer' applications.

16.8 Recommendations

I think it would be helpful if I translated my ruminations into tangible recommendations.

16.8.1 Programmatic Recommendations

Identify a set of technical goals that reflect European priorities. The American effort was driven primarily by NSA's interest in determining the feasibility of implementing Shor's algorithm and ARO and DARPA's interest in NP-complete problems and secure communications. In Europe, one might want a more focussed objective on some other computational or communications goals. I definitely think it helps coordinate a research programme if there are a set of grand challenges.

16.8.2 Technology Solutions

It is conceivable that a quantum computer architecture based on many simple quantum computing elements connected via a quantum network might be viable within five years. One view of such an architecture is as a quantum cellular automaton. Europe already has one of the best guys in the field for QCAs (Wim van Dam) so it would be a natural focus of attention. So, whereas the American strategy has been top-down (we want to factor 200 digit numbers, how do we do it?), the European strategy could be bottom-up (this is the architecture we can build today, what can we do with it that cannot be done classically?). In both cases, I believe it is important to have an over-arching mission-objective, e.g. a class of problems that you would like to solve, to guide the research directions and establish measurable milestones.

16.8.3 Quantum Computer Science

So far there are only a few quantum algorithms that tackle computational problems of practical significance. The field needs to find other 'killer applications'. What might these be? In an attempt to open up some new possibilities, Amir Fijany and I have developed quantum circuits for the quantum wavelet transform [8]. In classical computation, the wavelet transform is as important as the Fourier transform so there is a good chance that the quantum wavelet transform will be useful too.

Specific issues in quantum computer science that I would like to see explored are the automated discovery of new quantum algorithms [9], applications of the quantum wavelet transform [8], the development of probabilistic quantum algorithms that achieve non-unitary transformations some of the time, e.g., [10]. More generally, it would be helpful if we could move from the focus on coherence preservation to decoherence engineering. It is possible that decoherence could be harnessed rather than being a mere nuisance.

I believe it is also worthwhile to think about what one can do with just a few qubits. Certainly the quantum gyroscope seems like one possibility. More generally, one only needs a few qubits to perform certain physics simulations that have no known efficient classical algorithms, even though the ensemble quantum computing model is less powerful than standard quantum computing in the presence of oracles [11]. I would encourage research in this direction too as I believe it has the promise to give us intermediate goals on the road to full quantum computing.

16.8.4 Education and Training

I believe that we could, and should, teach quantum computing to undergraduates. I think the mathematical and conceptual demands are not so great, and this ultimately helps to seed the field with bright young people. Moreover, quantum computing offers a rather easy route into certain physics concepts such as superposition, interference, entanglement, non-clonability, and non-determinism. Regardless of whether a quantum computer can be built, the foundations of theoretical computer science have been forever changed and I believe that it is important to communicate this to our young people.

References

[1] C. Zalka, "Grover's Quantum Searching Algorithm is Optimal", http://xxx.lanl.gov/archive/quant-ph/9711070 (1997)
[3] J. P. Dowling, "Correlated Input-Port, Matter-Wave Interferometer: Quantum-Noise Limits to the Atom-Laser Gyroscope", Physical Review A, Volume 57, Number 6, June (1998)
[4] L. K. Grover, "Quantum Search on Structured Problems", to appear in Proceedings of the First NASA International Conference on Quantum Computing & Quantum Communications, Palm Springs, CA, Vol. 1509, Springer-Verlag Lecture Notes in Computer Science (1998) also available at http://www.lanl.gov/archive/quant-ph/9802035

[5] N. J. Cerf, L. K. Grover, C. P. Williams, "Nested Quantum Search and NP-complete Problems", http://www.lanl.gov/archive/quant-ph/9806078, (1998)

[6] D. S. Abrams and S. Lloyd, "Nonlinear Quantum Mechanics Implies Polynomial-Time Solution for NP-complete and #P Problems", http://xxx.lanl.gov/archive/quant-ph/9801041 (1998)

[7] R. Jozsa, "Quantum Effects in Algorithms", to appear in Proceedings of the First NASA International Conference on Quantum Computing & Quantum Communications, Palm Springs, CA, Vol. 1509, Springer-Verlag Lecture Notes in Computer Science (1998) also available at http://www.lanl.gov/archive/quant-ph/9805086

[8] A. Fijany and C. P. Williams, "Quantum Wavelet Transforms: Fast Algorithms and Complete Circuits", to appear in Proceedings of the First NASA International Conference on Quantum Computing & Quantum Communications, Palm Springs, CA, Vol. 1509, Springer-Verlag Lecture Notes in Computer Science (1998) also available at http://xxx.lanl.gov/archive/quant-ph/9809004

[9] C. P. Williams and A. Gray, "Automated Design of Quantum Circuits", to appear in Proceedings of the First NASA International Conference on Quantum Computing & Quantum Communications, Palm Springs, CA, Vol. 1509, Springer-Verlag Lecture Notes in Computer Science (1998)

[10] H. Bechmann-Pasquinucci, B. Huttner, and N. Gisin, "Nonlinear Quantum State Transformation of Spin-1/2, Phys.Lett. A242 (1998) 198-204 also at http://xxx.lanl.gov/archive/quant-ph/9708040

[11] E. Knill and R. Laflamme, "On the Power of One Bit of Quantum Information", http://xxx.lanl.gov/archive/quant-ph/9802037 (1998)

Section IV

Reference Materials

This section contains a research review (with a list of papers that cover the major areas and developments in QCC) and a breakdown of the field, including a glossary of frequently used terms.

Section IV

Reference Materials

Chapter 17
Quantum Information Processing: A Brief Overview of Recent Advances

Antonella Karlson
StarLab, Riverland nv

17.1 Introduction

In the mid-1930s two influential but seemingly unrelated papers were published. In 1935, Einstein, Podolsky and Rosen proposed the famous EPR paradox that has come to symbolize the mysteries of quantum mechanics. Two years later, Alan Turing introduced the universal Turing machine and laid the foundations of the computer industry. Although quantum physics is essential to understanding the operation of transistors and other solid state devices in computers, computation itself has remained a resolutely classical process. Surely the uncertainty associated with quantum theory is seemingly not compatible with the reliability expected from computers. In 1982, Richard Feynman suggested that individual quantum systems could be used for computations. In 1985, David Deutsch from the University of Oxford described the universal quantum computer and showed that quantum theory can allow computers to do more rather than less. An important new observation is that information is not independent of the physical laws which govern the system used to store and process it (Landauer). On the atomic scale matter obeys the laws of quantum mechanics, which are quite different from the ones of classical physics that determine the characteristics of conventional computers. Therefore quantum computers will have qualitatively new properties and capabilities. During the past ten years scientific groups all over the world have worked to establish the theoretical foundations and to investigate different experimental realizations of quantum computing and quantum communications.

On the other hand over the past 50 years computer technology has gone through a spectacular sequence of changes from one type of physical realization to another. Today's advanced lithographic techniques can place up to 7 million transistors per cubic centimeter of a silicon chip. Technology is being continuously improved and inevitably we will reach a point where logic gates are so small that they are made of a few atoms. Moore's law implies that we will be within the reach of atomic-scale computer devices within the next two decades. So if computers are to become smaller and faster in the future, new, quantum devices must replace or supplement classical ones.

17.2 The Underlying Physical System

17.2.1 Quantum Bits and Quantum Superpositions

One bit of information can be encoded using two different polarizations of light or two different electronic states of an atom or two different orientations of the spin of a quantum particle. In principle any quantum two state system can be used for this purpose and it is called a quantum bit or a **qubit**. Let us denote the two basis states with $|0>$ and $|1>$. While a classical bit can be set in either of these two states, there are an infinite number of possibilities for a qubit. It can be prepared also in any coherent superposition: $|\Psi> = (a\,|1> + b\,|0>)$ where $a^2 + b^2 = 1$. This is why one qubit can encode at a given moment of time both a logical 0 and 1. The numbers a and b are complex and can be specified to any precision required. Therefore, it takes an infinite amount of classical information to define the state. If we perform a measurement, the superposition is destroyed and the state reduces to one of the basis states $|0>$ and $|1>$. Therefore, any measurement of $|\Psi>$ will reveal at most one classical bit of information about its identity. We refer to the full (largely inaccessible) information represented by a quantum state as quantum information in contrast to the more familiar notion of classical information such as the outcome of a measurement which is, in principle, fully accessible.

One other specific of quantum states is the so called no cloning theorem: It is not possible to generate copies of a quantum state reliably, unless the state is already known.

Let us consider a system of a given number of interacting qubits, for example L spin-1/2 particles. The Hilbert space is constructed from all the possible states of the whole system. Therefore, there are 2^L basis states corresponding to 2^L possible bit strings with length L. Note that the dimensionality of the Hilbert space grows exponentially with L.

17.2.2 Quantum Gates

Quantum logic gates is a notion borrowed from classical information theory. It enables us to talk about quantum state manipulations without referring to the specific sequence of unitary transformations performed on the initial state. We only refer to the resultant state obtained from the initial one via a manipulation equivalent to a given logical operation.

It has been proven that all quantum one-bit gates and one two-bit gate, namely the quantum controlled-NOT gate, are sufficient to implement all logical operations on a QC. They can be used to construct arbitrary unitary transformations on any finite set of qubits.

Depending on the physical system, the unitary transformations can be realized in a different way by a time-dependent Hamiltonian. For example for a spin system it is a time-dependent magnetic field, laser fields are used to manipulate the discrete electronic levels of atoms or ions, optically active substances are used to change the polarization of photons.

17.2.3 Quantum Parallelism

Let us now make a register of L qubits which we assume are all initially in the state |0>. Then we perform the so-called Hadamard transformation on this register. $U_H |0\ 0\ 0\ ...\ 0> \rightarrow 2^{-L/2} (|0> + |1>)^L \rightarrow 2^{-L/2} \Sigma\ |k>$, where k is an integer represented as a binary number. The sum runs from zero to $2^L - 1$. Because of quantum superposition, the register can store 2^L different numbers at once. Therefore we obtain a superposition of every possible bit-string of length L. When the register is prepared in a superposition of 2^L different numbers, we can perform operations on all of them simultaneously. During such evolution each state in the superposition is affected and the outcome is the same superposition of the manipulated states. As a result we generate a massive parallel computation in one piece of quantum hardware. This means that a quantum computer can in only one computational step perform the same mathematical operation on 2^L different input numbers encoded in coherent superpositions of L qubits. In order to accomplish the same task any classical computer has to repeat the same computation 2^L times or one has to use 2^L different processors working in parallel. In other words, a quantum computer offers an enormous gain in the use of computational resources such as time and memory. The point is that to solve the same problem classical computers need an amount of time or memory which is exponentially larger than the ones quantum computers need.

17.3 Fundamentals of Quantum Information

17.3.1 Entanglement

Entanglement is one of the distinct properties of quantum systems (together with quantum superposition and probabilistic measurement among others) that makes quantum information processing so different from classical information technology. This phenomenon refers to the joint state of two or more quantum systems and describes correlations between them that are much stronger than any classical correlations. In technical terms an entangled state is any state that cannot be written as a direct product of one particle states (in any basis). Entangled states offer the possibility to encode information in a completely new way. Let us assume we have two qubits and we want to encode two bits of information. The straightforward approach is to encode one bit of information onto each qubit separately: $|0>_1|0>_2$, $|0>_1|1>_2$, $|1>_1|0>_2$ and $|1>_1|1>_2$. But using entangled states it is possible to do it in such a way that neither of the two qubits carries any well defined information on its own: all the information is encoded in their joint properties. A widely used choice of entangled states are the so-called **Bell states**: $|\Psi^+> = 1/\sqrt{2}\ (|0>_1|1>_2 + |1>_1|0>_2)$, $|\Psi^-> = 1/\sqrt{2}\ (|0>_1|1>_2 - |1>_1|0>_2)$, $|\Phi^+> = 1/\sqrt{2}\ (|0>_1|0>_2 + |1>_1|1>_2)$, $|\Phi^-> = 1/\sqrt{2}\ (|0>_1|0>_2 - |1>_1|1>_2)$. Entanglement makes possible quantum teleportation, quantum error correction, quantum dense coding, etc.

Entanglement is closely linked to the issue of **non-locality** in quantum theory. Assume that two particles in an entangled state are widely separated. If we perform a measurement on one of them, this will immediately influence the quantum state of the other one. So we know in advance what the state of the second particle is without actually measuring it. It seems as if the particles are communicating faster than the speed of light, but special relativity is not violated because no information is exchanged.

The basis chosen to encode quantum information and the states chosen to represent $|0>$ and $|1>$ are arbitrary. Let us assume we have photons whose polarization we are measuring in a chosen basis. We identify the horizontal polarization in this basis as '0' and the vertical - as '1'. We can now choose a new 'conjugate' basis, which is rotated with respect to the first one at 45° in the plane of polarization. The states in the new basis are related to the previous ones by a rotation in Hilbert space (known as the Hadamard transformation): $|0'> = 1/\sqrt{2}\ (|0> + |1>)$ and $|1'> = 1/\sqrt{2}\ (|0> - |1>)$. Conjugate bases cannot be used at the same time in an experiment. But the possibility to switch alternatively between various bases is the foundation of many useful methods.

It is possible to have entanglement in higher dimensions. For example the so-called Greenberger-Horne-Zeilinger (GHZ) states are entangled superpositions of three qubits.

17.3.2 Quantum Dense Coding

Entanglement permits a new way of encoding information (Bennett, Wiesner). Note that it is possible to obtain any one of the four Bell states from any other by performing only one of the following four unitary operations on just one of the two qubits: identity; phase shift; bit flipping; phase shift and bit flipping.

Consider Alice and Bob who are far way from each other. Bob can send two bits of information to Alice with a single photon if they share a pair of entangled photons. To begin, one photon each is sent to Alice and Bob. The photons are in one of four Bell states. Bob then performs one of the above mentioned four unitary transformations on his photon. That transfers the pair into another Bell state or simply does nothing. Bob then sends his photon to Alice who measures the state of the pair. Since there are four possible outcomes of the measurement, Bob has sent twice as much information as can be sent classically with a two-state particle. This scheme has been experimentally realized by the group of Zeilinger in Innsbruck using polarization-entangled photons.

17.3.3 Quantum Teleportation

Suppose Alice wants to teleport a photon in an unknown quantum state $|\Psi\rangle = (a\,|0\rangle + b\,|1\rangle)$ to Bob. They both agree to share an entangled pair of qubits, known as the **ancillary pair**. For example, let it be in the state $|00\rangle + |11\rangle$. The initial state of the three qubits is then $(a\,|000\rangle + b\,|100\rangle + a\,|011\rangle + b\,|111\rangle)$. Alice now measures in the Bell basis the first two qubits, i.e. the unknown one and her member of the entangled pair. The network to do this consists of a controlled-NOT gate on the first two bits and consecutively an Hadamard gate on the first one. Just before the measurement her qubits are in the state: $|00\rangle\,(a|0\rangle + b|1\rangle) + |01\rangle\,(a|1\rangle + b|0\rangle) + |10\rangle\,(a|0\rangle - b|1\rangle) + |11\rangle\,(a|1\rangle - b|0\rangle)$. The measurement of Alice collapses the state of the third qubit into one of four possibilities and yields two classical bits. The latter are sent classically to Bob who uses them to learn which of the four unitary operators he has to apply to his ancillary bit in order to place it in the state $|\Psi\rangle$. Thus, Bob ends up with the qubit (**the quantum information and not the actual quantum system**) which Alice wished to transmit. During the whole process the state of the teleported photon remains unknown. This means that **any quantum state can be teleported**. Note that quantum information can arrive to Bob only if it disappears from Alice (no cloning). Also, **quantum information is complete information**: $|\Psi\rangle$ is the complete description of Alice's qubit. Relativity remains intact because two classical bits of information have to be transmitted in order for Bob to obtain the state $|\Psi\rangle$.

This teleportation experiment was performed by Zeilinger's group at the University of Innsbruck and a related experiment was carried out at the University of Rome.

17.4 Quantum Cryptography

17.4.1 Standard Cryptosystems

Cryptography consists of hiding information in a string of bits that are meaningless to any unauthorized party. An algorithm is used to combine a message with some additional information, known as the **key**, to produce a cryptogram. This technique is known as **encryption**. The person who encrypts and transmits the message is traditionally known as Alice. The person who receives the message is known as Bob. He decrypts the scrambled message using his key. Eve is an unauthorized eavesdropper. For a cryptosystem to be secure, it should be impossible to unlock the cryptogram without Bob's key. In practice, this demand is often softened so that the system is just extremely difficult to crack.

The **'one-time pad'** cryptosystem was proposed by Vernan in 1935 and allows messages to be sent with perfect security. Alice chooses a random bit string for the key which must be at least as long as the message itself. She encrypts her message by adding the key to it (calculations are performed in modulo 2). She then transmits the scrambled message to Bob, who decrypts it by subtracting the key. The problem with this system is that both Alice and Bob have to share the key. It has to be transmitted by some trusted means. The key can be used only one time.

Another class of cryptosystems shares a **public key**. They are based on the so-called **one way functions**: it is easy to compute $f(x)$ knowing x and difficult in the opposite direction. Here difficult means that the time for the computation grows exponentially with the size of the input. Factoring large integers is an example of such function and the **RSA** cryptosystem is based on it. For Alice to transmit a message with a public key cryptosystem, Bob first chooses a private key. He uses this key to compute a public key which he discloses publicly. Alice then uses the public key to encrypt her message. She transmits it to Bob, who decrypts it with his private key. The encryption-decryption process can be described mathematically as a one-way function with a trapdoor - the private key. This system has two major drawbacks. First, nobody knows if an efficient classical algorithm for factorization exists. Second, the current advances in quantum computing show that factorization can be performed efficiently if these machines are built in the future. Therefore the RSA system is threatened to become obsolete.

17.4.2 Quantum Key Distribution

Quantum cryptography allows two physically separated parties to **create a random secret key** without resorting to the services of a courier. It also allows them to verify whether the key has been intercepted. When used with the one-time pad scheme, the key allows the message to be transmitted with proven and absolute security. Therefore quantum key distribution is really a better name.

Let us consider the 'BB84' protocol, introduced by Bennett and Brassard in 1984. Alice and Bob are widely separated and wish to communicate. Alice sends to Bob $2n$ qubits, each prepared in one of the states: $|0>$, $|1>$ and $|\pm> = 1/\sqrt{2}$ ($|1> \pm |0>$) randomly chosen. Bob measures the received bits, choosing the measurement basis randomly between $|0>$, $|1>$ and $|+>$, $|->$. Next, Alice and Bob inform each other publicly of the basis they used to prepare or measure each qubit. They find out on which occasions they by chance used the same basis (which happens on the average half of the time) and retain just those qubits. In the absence of errors or interference, they now share the same random string of n classical bits. This classical bit string is called the **raw quantum transmission RQT**.

The important feature is that it is impossible for anyone to learn Bob's measurement results by observing the qubits during transmission, without leaving evidence of their presence. Eve can attempt to discover the key by intercepting the qubits, measuring them and then passing them on to Bob. On average, half the time Eve guesses the basis of Alice correctly and thus does not disturb the qubit. However, Eve's correct guesses do not coincide with Bob's. So Eve learns the state of half of the n qubits which Alice and Bob later decide to trust and disturbs the other half (for example sending to Bob $|+>$ instead of Alice's $|0>$). Half of those disturbed qubits will be projected by Bob's measurement back onto the original state sent by Alice. Therefore Eve corrupts $n/4$ bits of the RQT.

Alice and Bob can now detect Eve's presence simply by randomly choosing $n/2$ bits of the RQT and announcing publicly the values they have. If they agree on all these bits, they can trust that no eavesdropper was present. The $n/2$ undisclosed bits form the secret key. In practice the protocol is more complicated since noise will corrupt some of the qubits. Alice and Bob retain the key as long as they find the error rate to be well below 25%. After detecting and removing the errors, they are left with a key of around $n/4$ bits.

Another protocol, proposed in 1991 by Artur Ekert from the Center for Quantum Computation at Oxford University, involves the use of EPR pairs, which Alice and Bob measure along one of three different axes. To rule out eavesdropping, they check for Bell-EPR correlations in their result.

Quantum key distribution is feasible with current technology. Nicolas Gisin's group at the University of Geneva demonstrated in 1995 the so-called polarization encoding scheme of photons below lake Geneva. BT, Los Alamos National Laboratory in the US and IBM are other major players in this field.

17.5 Quantum Computing

17.5.1 Quantum Algorithms

Quantum mechanics allows information processing of a kind which is ruled out by classical physics. As a result quantum computers can support entirely new kind of computations with qualitatively new algorithms which are based on the quantum properties of the underlying physical system and do not have classical analogues. They are called quantum algorithms and make use of the possibility to simultaneously manipulate coherent quantum superpositions of input states (quantum parallelism). This results in the possibility to change the complexity of certain problems.

A fundamental notion in the theory of computational complexity is the distinction between polynomial and exponential use of resources in computation. This provides a quantitative measure of distinction between classical and QC. Let the input for a computational task be the number N. We wish to assess the resources required for this task as a function of the size of the input which is $L = log_2 N$, i.e. the number of the bits needed to store N. If $T(L)$ denotes the number of steps needed to solve the problem (on a standard universal computer), we ask whether $T(L)$ can be bounded by some polynomial function in L or whether $T(L)$

grows faster than any polynomial. The problem is said to be in complexity class *P* if there exists an algorithm which runs in time *T(L)* bounded by a polynomial function. Otherwise the problem is said to require exponential time. The term 'efficient computation' is also used for problems which run in polynomial time. The exact number of steps will depend on the choice of the computer and the model of computation. But if we stay within the framework of classical physics, the distinction between polynomial and exponential time is robust and is independent of these choices. From the physical point of view it is natural to extend the notion of efficient computation to require the efficient use of all possible physical resources (not only time) like memory, energy, etc. The fundamental *raison d'être* of QC is the fact that it appears to allow transgression of the classical boundary between polynomial and exponential computations.

The theory of computation and computational complexity was normally presented as an entirely mathematical theory with no reference to considerations of physics. It was believed that the answer to the question whether any given problem could be solved in a time that was polynomial in the size of the input, or greater than polynomial, was independent of the physical apparatus used to perform the computations. This is indeed true for all computers based on classical physics. However any actual computation is a physical process involving the physical evolution of selected properties of the physical system. Consequently the issues of 'what is computable' and 'what is the complexity of the computation' must depend essentially on the laws of physics and cannot be characterized by mathematics alone. The formalism of quantum physics allows enlargement of the boundaries of the classical theory of computational complexity, whose formulation was based implicitly on classical physics. As things are now, quantum processes cannot result in any computation which is not already possible by classical means, but there is a significant difference in the efficiency of computation.

It is not yet clear what is the essential nature of quantum algorithms, what is the general class of computational problems that can be solved efficiently on a QC and whether a whole set of quantum algorithms exist.

17.5.2 Grover's Search Algorithm

In 1997, Grover, from AT&T Bell Labs in the US described a quantum algorithm for searching a random database: given a list of items, find a particular one of them. For example, find a particular telephone number in a telephone directory (for someone whose name you do not know). For a list of *N* items, classical algorithms require on the average *N/2* steps. Grover's algorithms requires $\sqrt{N}$ steps. The problem is not transferred to a new complexity class, but it is speeded up significantly. Bennett proved that Grover's algorithm is optimal: no quantum algorithm can do better than $O(\sqrt{N})$.

The algorithm proceeds as follows:

- Start with a L qubit register at |0>.

- Apply the Hadamard transform on the register: $U_H |0\ 0\ 0\ ...0\rangle = 2^{-L/2} (|0\rangle + |1\rangle)^L = 2^{-L/2} \Sigma\ |x\rangle$ where the sum runs from zero to 2^L-1
- Apply the operator U_v defined by: $U_v |x\rangle = |x\rangle$ for $x \neq v$ and $U_v |v\rangle = -|v\rangle$ where v is the item we are looking for.
- Apply the gate $U_D = U_H U_0 U_H$ where $U_0 |0\rangle = -|0\rangle$ and $U_0 |x\rangle = |x\rangle$ for $x \neq 0$.
- Repeat the last two steps k times.

At the end of the first iteration the amplitude of the state $|v\rangle$ has become slightly larger than the amplitude of the rest of the states in the superposition. It can be shown that after each iteration the state can be written as $|\Psi\rangle = \Sigma_{x \neq v}\ a\ |x\rangle + a_v|v\rangle$. Therefore the quantum superposition is evenly weighted, apart from the state $|v\rangle$ whose amplitude departs from the amplitude of the rest of the states as the algorithm is iterated. Boyer and co-workers derived an analytical expression for each iteration: $a_v(k) = sin((2k+1)\theta)$ where $sin^2(\theta) = 2^{-L}$. As the number of iterations k increases, $a_v(k)$ increases as well and it reaches a maximum at $k_{max} = \frac{1}{4} \pi\ 2^{L/2}$ iterations.

Grover's algorithm can be extended when more than one element $|v\rangle$ satisfies the condition. Variations of this algorithm can also find the largest or smallest value in a list, approximate the mean value, etc. It is also an important tool in cryptoanalysis. Classical cryptographic schemes can be cracked by searching among 10^{16} possible keys. A classical computer would take over a thousand years, while a QC using Grover's algorithm would take around four minutes. Therefore it is a very versatile searching tool.

17.5.3 Period Finding and Shor's Factorization Algorithm

In 1993, Shor from AT&T Bell Labs in the US discovered a quantum algorithm that can factor large integers in polynomial time. Shor used the fact that the factorization problem can be related to finding periods of certain functions. In particular, it can be shown that finding the two prime factors p and q of $N = pq$ can be reduced to finding the period of the function $f_{a,N}(x) = a^x \mod N$. Here a is any randomly chosen number smaller than N which is coprime with N, i.e. which has no common factors with N. (If a is not coprime with N, then the factors of N are trivially found by computing the greatest common divisor $gcd(a,N)$ using the Euclidean algorithm.) This function gives the remainder after the division of a^x by N. The function is periodic with period $r = r(a,N)$.

If we know the period r, we can factor N provided r is even and $a^{r/2} \mod N \neq -1$. When a is chosen randomly between 0 and N, these two conditions are satisfied with probability greater than 1/2. The factors of N are then given by the $gcd(a^{r/2} \pm 1, N)$. For this calculation exists a very efficient algorithm since 300 BC, namely the Euclidean algorithm which runs in polynomial time. Thus the problem of factoring big numbers reduces to finding the period of a function. Every step of this method, except finding r, can be performed in polynomial time on a classical computer. Unfortunately, finding *r* is as time consuming as finding the factors of N. Shor describes a randomized quantum algorithm which provides the period *r* of a function and runs efficiently on a quantum computer.

17.5.4 Minimum Requirements for any Quantum System to Be a Quantum Computer

1. The degrees of freedom required to hold data and perform computation should be available as dimensions of the Hilbert space of the quantum system. This Hilbert space should be precisely enumerable and only statistical statements about the number of degrees of freedom is not sufficient. It is also essential that the size of the Hilbert space grows exponentially with the size of the system. This can be achieved with a multi-particle quantum system.
2. It must be possible to place the system in a given well defined quantum state. The latter can be very simple, like "all spins down" for an ensemble of spins. This can be satisfied if it is possible to cool the system to its ground state.
3. The system must be to a high degree isolated from coupling to the environment. This minimizes correlations with the environment and therefore also decoherence.
4. It must be possible to subject the system to a controlled sequence of unitary transformations. All quantum algorithms are expressed in terms of such sequences. These transformations must be made to act upon specified pairs of qubits or other small collections of qubits. Quantum logic gates produce entanglement between qubits.
5. It is possible to subject the system to measurements conform with quantum mechanics.

This formulation was done by DiVincenzo from IBM in the US.

17.5.5 Simulation of Other Quantum Systems

The first proposed application of quantum computers was to a class of problems in physics: in 1982, Richard Feynman conjectured that quantum computers could be much more efficient than classical computers for the problem of simulating other quantum systems. Simulating quantum systems, e.g., by Monte Carlo techniques, is a difficult task that uses large amounts of computer time. The basic problem is that a quantum system with N degrees of freedom occupies a Hilbert space of size at least 2^N (even writing down the state of a quantum system requires an exponentially large amount of space on a classical computer) and evolving that state forward in time requires the exponentiation of matrices of dimension 2^N. As a result, following the time evolution of a quantum system of, say, 40 spins is effectively impossible on today's computers and on computers for some time to come. In 1996, Lloyd showed that Feynman's conjecture was correct: quantum computers could function as universal quantum simulators, providing efficient analogue simulations of any quantum system that evolves according to local interactions. These results were subsequently applied to fermionic systems, spin glasses, to molecular and atomic wavefunctions and spectra, and to quantum chaos: for such systems, a quantum computer with a few tens of qubits could, in principle, provide more accurate simulations than current classical computers.

In 1998, David Cory provided an experimental demonstration of the power of quantum simulation: using nuclear magnetic resonance techniques, he made a sample of calcium fluoride containing approximately 10^{12} spins perform a quantum simulation of a spin system with long-range spin-flip interactions not known to occur in nature. Cory used this quantum simulator to perform a quantum calculation of the spin diffusion rate of the artificial system, obtaining a result compatible with semi-classical mean field approximations of this rate. To perform the same calculation on a classical computer would require a computer with at least 2 to the power 10^{12} elements, a number much greater than the number of elementary particles in the universe. Cory's technique is not amenable to simulation of arbitrary quantum systems, but a substantial set of systems can be addressed by these methods. Once 'general purpose' quantum computers can be constructed with more than 30 or so bits, calculations of the properties of a large variety of quantum systems should become more accurate on quantum computers than on classical devices.

17.6 Quantum Decoherence

17.6.1 What is Decoherence?

In principle we know how to build a quantum computer. We can start with simple quantum logic gates and try to integrate them together into quantum circuits. However, if we keep on with this process, we will quickly run into some serious practical problems. Apart from the technical difficulties of working at single-atom and single-photon scales, one of the biggest obstacles is quantum decoherence. If the system is not perfectly isolated from its environment, the quantum dynamics of the surrounding apparatus will also be relevant to the operation of the QC. Its effect will be to make the computers evolution nonunitary, because it is not a closed system any more. The slightest outside disruption (heat or light for example) can destroy the balance of the quantum states that store the information and this would make the computation impossible. Even the very process of measuring the state of a qubit can destroy the coherence. It was shown that decoherence leads to an exponential increase of the error rate with the input size. Therefore, a given quantum algorithm cannot be considered as efficient any more, regardless of how weak the interaction with the environment may be.

A classical computation follows a single definite pathway in time from the beginning to the end. In QC the computation can be split up into several pathways that, by the principle of superposition, evolve in time in parallel. These pathways recombine at the end and because each of them carries a definite phase, they can interfere constructively or destructively. Loss of phase coherence will destroy the interference. Therefore the coherence time τ_c needs to be longer than the running time of the computation. Continuous advances in the experimental techniques result in increasing τ_c. The minimum time required to execute one quantum gate is denoted by τ_s and is called clock cycle. It is equal to the time needed to flip a qubit and it is estimated as $\tau_s = \hbar/\Delta(E)$, where $\Delta(E)$ is the typical energy splitting in the two-level system. The ratio $m = \tau_c / \tau_s$ is equal to the largest number of steps permitted in a QC using these qubits.

In the table below are given the characteristic times (in seconds) for various two-level systems that might be used as quantum bits:

qubit	τ_s	τ_c	m
Electrons: GaAs	10^{-13}	10^{-10}	10^3
Electrons: Au	10^{-14}	10^{-8}	10^6
Trapped ions	10^{-14}	10^{-1}	10^{13}
Optical microcavity	10^{-14}	10^{-5}	10^9
Electron spin	10^{-7}	10^{-3}	10^4
Electron quantum dot	10^{-6}	10^{-3}	10^3
Nuclear spin	10^{-3}	10^4	10^7

17.6.2 Quantum Error Correction

Quantum error correction shows how the QC can survive interactions with the environment and how to tolerate slightly inaccurate quantum gate operations. It was put into a formal framework in 1995 by Peter Shor and independently by Andrew Steane from Oxford University. They proposed **error correcting codes** which allow a physical process previously thought to be impossible. Namely a general quantum state can be almost perfectly recovered, undoing even irreversible processes such as relaxation by spontaneous emission. During a long error corrected quantum computation, using fault-tolerant methods, every qubit in the computer might decay a million times and yet the coherence of the quantum information is preserved. All the operations performing the error correction can be realized in a way so that they introduce less noise than they can suppress.

The basic idea is rather simple: the unknown state of a two-level system is encoded into an entangled state of several qubits. If any (unknown) qubit is arbitrarily corrupted, then the original state can be perfectly reconstructed. None of the information of the original state resides locally in the encoding. In this way the state may be protected against the effects of environmental interactions if these interactions act by local means.

Error correction requires ten to hundred times increase of the number of qubits in a QC and thousands of extra quantum gates, but it overcomes a major hurdle. The claim which is presently being evaluated is that if the error rate (induced by interaction with the environment, as well as by inaccuracies in the implementation of the unitary transformations) is below a certain value, then reliable QC of indefinite scale and duration become possible. This error rate threshold of reliable QC is presently very low, about 10^{-5} per qubit per clock cycle.

17.7 Experimental Realizations

The current challenge is not to build a QC right away, but rather to move from experiments in which we observe quantum phenomena to experiments in which we can control these phenomena. QC requires the preparation, manipulation, coherent evolution and measurement of a pure quantum state.

17.7.1 Systems with Few Degrees of Freedom

Linear ion traps are one of the most promising realizations of a QC. The ground state of an ion can represent a logical 0, while any other excited state a 1. Transitions between the two levels can be induced by a laser pulse of a specific intensity, duration and frequency. The ions are placed along a string in vacuum. They are trapped with electromagnetic fields of appropriate configuration and cooled down with a set of laser beams to micro Kelvin temperatures. The ions interact with each other exchanging vibrational excitations. Each ion can be separately controlled by a properly focused and polarized laser beam. The Hilbert space is spanned by the energy levels of the isolated ions, combined with the phonon modes of the ions in the trap. The coherence times of the ion levels is unmeasurably long. The phonon lifetimes are adequately long. Quantum gates can be performed with very high precision by laser spectroscopy techniques.

In 1995, Ignacio Cirac and Peter Zoller of the University of Innsbruck showed how to implement the controlled-NOT operation. In the last few years the group of Christopher Monroe and David Wineland from NIST in Boulder, USA have improved these techniques.

In **cavity quantum electrodynamics** experiments a strong coupling can be achieved between a single atom or ion and a single mode of the electromagnetic field in the cavity. This coupling can be used to apply quantum gates between the field mode and the ion, thus making possible to transfer quantum information between separated ion traps via optical cavities and fibers. Leading groups are Serge Haroche and Jean-Michel Raimond at Ecole Normale Superieure in Paris and independently at Caltech in Pasadena.

17.7.2 Macroscopic Systems

In 1997, a completely different approach to the practical realization of QC was proposed by N. Gershenfeld and I. Chuang from MIT and, independently, by D. Cory, A. Fahmy and T. Havel from Harvard. It is based on using macroscopic ensembles of identical known molecules in a liquid, instead of isolated quantum particles. A number of nuclear spins in each of these molecules are used as qubits. They are manipulated with the techniques of modern NMR spectroscopy. Spin-spin interactions cause entanglement between the qubits on each molecule. A perturbation into the system's density matrix is introduced. It acts like a low dimensional effective pure state. The initial state of the spins is the input data. Then radio frequency pulses perform unitary transformations onto the spin system, which is the actual computation. This model differs from (conventional) QC only in the fact that the result of a measurement is the expectation value of an observable rather than a random eigenvalue of it. The biggest advantage of this approach is that, because of the massive redundancy provided by having a large ensemble of identical copies of the system, environmental interactions or intentional measurements only weakly perturb the computer's state. Therefore decoherence is insignificant. At present there are three experimental groups (D. Cory and T. Havel at Harvard, J. Jones and A. Ekert at the Center for Quantum Computation at Oxford University and I. Chuang and N. Gershenfeld at MIT) who have implemented Deutsch's algorithm and recently Grover's algorithm with two qubits. However, the experimental techniques available today limit this QC realization to 6 - 8 qubits.

17.8 Conclusions

- If large QC are ever made they will be used to address just those specific tasks which benefit from quantum information processing.

- Quantum error correction has shown how to overcome the main obstacles to the implementation of QC. Therefore, there are no fundamental reasons against the practical realization of QC.

- The ideas of classical information theory fit into quantum mechanics.

- Quantum information is independent of the details of the quantum system in which it is stored.

- The very concept of information has to be changed. The foundations of computer science have to be reformulated on the basis of quantum physics instead of mathematics or classical physics.

- The field of QCC leads to a series of practical applications like quantum communication, cryptography, simulations and computing. On the other hand it brings a better and deeper understanding of many fundamental concepts of the quantum theory and of the physical world.

Further Reading

Introduction: Bits, Superposition, Gates, and Parallelism
R.P. Feynman, 'Simulating Physics with Computers,' *International Journal of Theoretical Physics*, **21**, 467-488 (1982).
D. Deutsch, 'Quantum Theory, the Church-Turing Principle and the Universal Quantum Computer', *Proc. Royal Soc. of London* **A 400**, 97-117 (1985).
Steane, 'Quantum Computing', eprint quant-ph/9708022 (http://xxx.lanl.gov/archives/quant-ph).
Barenco, A. Ekert, A. Sanpera, C. Macchiavelo, 'A Short Introduction to Quantum Computation', http:/www.qubit.org/intros/comp/comp.html
S. Lloyd, 'Quantum-Mechanical Computers,' *Scientific American*, **273**, 140-145 (1995).
D.P. DiVincenzo, 'Quantum Computation,' *Science* **270**, 255-261 (1995).
S. Braunstein, 'Quantum Computation: a Tutorial', http://chemphys.weizmann.ac.il/~schmuel/ comp/comp.html
G. Brassard, 'New Trends in Quantum Computing', eprint quant-ph/9602014.
Berthiaume, 'Quantum Computation', http://andre.cs.depaul.edu/andre
J. Preskill, 'Quantum Computing: Pro and Con', eprint quant-ph/970532
D. P. DiVincenzo, 'Topics in Quantum Computers', eprint quant-ph/9612126.

Information, Entanglement, Dense Coding and Teleportation
A. Zeilinger, 'Fundamentals of Quantum Information', *Physics World* **11**, 35-40 (1998).
D. Deutsch, A. Ekert, 'Quantum Communication Moves to the Unknown', http://www.qubit.org/intros/comm/comm.html
C.H. Bennett, G. Brassard, C. Crepeau, R. Jozsa, A. Peres, W.K. Wootters, 'Teleporting an Unknown Quantum State via Dual Classic and EPR Channels', *Phys. Rev. Lett.* **70**, 1895-1899 (1993).
C.H. Bennett, S.J. Weisner, 'Communication via One and Two Particle Operators on EPR States', *Phys. Rev. Lett.* **69** 2881-2884 (1992).
D. Bouwmeester, J-P Wan, K. Mattle, M. Eibl, H. Weinfurter, A. Zeilinger, 'Experimental Quantum Teleportation', *Nature* **390**, 575-579 (1997).
D.M. Greenberger, M.A. Horne, A. Shimony, A. Zeilinger, 'Bell's Theorem Without Inequalities', *Am. J. Phys.* **58**, 1131-1143 (1990).
D.M. Greenberger, M.A. Horne, A. Zeilinger, 'Multi-particle Interferometry and the Superposition Principle', *Physics Today*, 22-29, August 1993.

E. Hagley et al, 'Generation of EPR Pairs of Atoms', *Phys. Rev. Lett.* **79**, 1-5 (1997).
K. Mattle, H. Weinfurter, A. Zeilinger, 'Dense Coding in Experimental Quantum Communication', *Phys. Rev. Lett.* **76**, 4656-4659 (1996).

Aspects of Cryptography
W. Tittel, G. Ribordy, N. Gisin, 'Quantum Cryptography', *Physics World* **11**, 41-45 (1998).
Ekert, 'What is Quantum Cryptography?', http://www.qubit.org/intros/crypt.html
Ekert, 'From Quantum Code-Making to Quantum Code-Breaking', eprint quant-ph/-9703035.
A. Ekert, 'Quantum Cryptography Based on Bell's Theorem', *Phys. Rev. Lett.* **67** 661-663 (1991).
A. Fuchs, 'Optimal Eavesdropping in Quantum Cryptography', *Phys. Rev.* **A 56** 1163-1172 (1997).
R. J. Hughes, 'Quantum Cryptography', *Contemp. Phys.* **36** 149-163 (1995).
S.J.D. Phoenix, P.D. Townsend, 'Quantum Cryptography: How to Beat the Code Breakers Using Quantum Mechanics', *Contemp. Phys.* **36** 165-195 (1995).

Quantum Computing and Algorithms
D. Deutsch, A. Ekert, 'Quantum Computation', *Physics World* **11**, 47-52 (1998).
Deutsch and R. Jozsa, 'Rapid Solution of Problems by Quantum Computation,' *Proceedings of the Royal Society of London* **439**, 553-558 (1992).
D.R. Simon, 'On the Power of Quantum Computation,' in *Proceedings of the 35th Annual Symposium on Foundations of Computer Science*, S. Goldwasser, Ed., IEEE Computer Society, Los Alamitos, CA, 1994, 116-123.
R. Cleve, A. Ekert, C. Macchiavello, and M. Mosca, 'Quantum Algorithms Revisited', *Proc. R. Soc. Lond.* **A** 454, 339-354 (1998).
L.K. Grover, 'Quantum Mechanics Helps in Searching for a Needle in a Haystack,' *Physical Review Letters* **79**, 325-328 (1997).
M. Boyer, G. Brassard, P. Hoyer, A. Tapp, 'Tight Bounds on Quantum Searching', eprint quant-ph/9605034.
L.K. Grover, J. Smolin and B. Terhal, to appear eprint quant-ph 1998.
M. Mosca, 'Quantum Searching, Counting and Amplitude Amplification by Eigenvector Analysis', to be published.
P. Shor, 'Algorithms for Quantum Computation: Discrete Log and Factoring,' in *Proceedings of the 35th Annual Symposium on Foundations of Computer Science*, S. Goldwasser, Ed., IEEE Computer Society, Los Alamitos, CA, 1994, 124-134.
Ekert, R. Jozsa, 'Quantum Computation and Shor's Factoring Algorithm', *Rev. Mod. Phys.* **68** 733-753 (1996).
R. Cleve, A. Ekert, C. Macchiavello, M. Mosca, 'Quantum Algorithms Revisited', eprint quant-ph/9708016.

Requirements for Quantum Computers
D. P. DiVincenzo and D. Loss, 'Quantum Information is Physical', *Superlattices and Microstructures* **23** 419-432 (1998) and eprint cond-mat/9710259.
DiVincenzo, 'Topics in Quantum Computers', e-print cond-mat/9612126 v2.

Quantum Simulators
S. Lloyd, 'Universal Quantum Simulators,' Science **273**, 1073-1078 (1996).
S. Wiesner, 'Simulations of Many-Body Quantum Systems by a Quantum Computer', eprint quant-ph/9603028.
Abrams, S. Lloyd, 'Simulation of Many-Body Fermi Systems on a Universal Quantum Computer,' *Physical Review Letters* **79**, 2586-2589, 1997.
D. Lidar, to appear in *Phys. Rev. A*.
D. Abrams, S. Lloyd, to be published.
R. Schack, 'Using a Quantum Computer to Investigate Quantum Chaos', eprint quant-ph/9705016.
D. Cory, to be published.

Decoherence and Error Correction
D. DiVincenzo, B. Terhal, 'Decoherence: the Obstacle to Quantum Computation', *Physics World* **11**, 53-57 (1998).
M. Palma K-A Suominen, A. Ekert, 'Quantum Computers and Dissipation', eprint quant-ph/9702001.
A. Steane, 'Multiple Particle Interference and Quantum Error Correction', eprint quant-ph/9601029.
Ekert, Ch. Macchiavello, 'Quantum Error Correction for Communication', eprint quant-ph/9602022.
P. Shor, 'Fault-Tolerant Quantum Computation', eprint quant-ph/9605011.
D.P. DiVincenzo, P.W. Shor, 'Fault-Tolerant Error Correction with Efficient Quantum Codes', *Phys. Rev. Lett.* **77** 3260 (1996).
D. Gottesman, 'Stabilizer Codes and Quantum Error Correction', *Ph.D. Thesis Caltech*, eprint quant-ph/9705052.
C.J. van Enk, J.I. Cirac, P. Zoller, 'Ideal Quantum Communication over Noisy Channels: a Quantum Optical Implementation', *Phys. Rev. Lett.* **78** 4293 (1997).

Experimental Realizations
A. Steane, 'The Ion Trap Information Processor', eprint quant-ph/9608011.
D.J. Wineland *et al*, 'Experimental Issues in Coherent Quantum State Manipulation of Trapped Atomic Ions', eprint quant-ph/9710025.

Macroscopic Systems
D. Cory, A. Fahmy, T. Havel, 'Ensemble Quantum Computing by NMR Spectroscopy', *Proc. Natl. Acad. Sci. USA* **94**, 1634-1639 (1997).
D. Cory, M. Price, A. Fahmy, T. Havel, 'NMR Spectroscopy: an Experimentally Accessible Paradigm for Quantum Computing',
eprint quant-ph/9709001.

N. Gershenfeld, I. Chuang, 'Bulk Spin-Resonance Quantum Computation', *Science* **275** 350 (1997).

J. A. Jones and M. Mosca 'Implementation of a Quantum Algorithm on a Nuclear Magnetic Resonance Quantum Computer', *J. Chem. Phys.* **109** 1648-1653 (1998) and quant-ph/9801027.

J.A. Jones, M. Mosca, R. Hansen, 'Implementation of a Quantum Search Algorithm on a NMR QC', *Nature*, **393**, 344-346 (1998).

J.A. Jones, R.H. Hansen, M. Mosca, 'Quantum Logic Gates and NMR Pulse Sequences', quant-ph/9805070.

J. A. Jones, 'Fast Searches with Nuclear Magnetic Resonance Computers', *Science* **280**, 229 (1998).

J.A. Jones, 'Quantum Computing and NMR' in *The Physics of Quantum Information*, edited by D. Bouwmeester, A. Ekert, and A. Zeilinger, Springer-Verlag, Berlin, in preparation.

J. A. Jones and M. Mosca, 'Approximate Quantum Counting on an NMR Ensemble Quantum Computer', submitted to *Physical Review Letters*, also quant-ph/9808056.

Chapter 18
Categories and Definitions

18.1 Definitions and Glossary of Terms

The following were suggested as part of the Pathfinder project's goal of a preliminary analysis of this field. However, the field is constantly evolving, and the subdivision of the various research areas may prove to be inadequate or even incorrect in the not too distant future. Thus these definitions may soon seem incomplete: they are included here only as a guide to the subject areas within this somewhat amorphous field.

18.1.1 Definition of Quantum Computing and Communications

Quantum Computing and Communications (QCC) aims to apply specific aspects of quantum theory in the development of new systems and techniques for information processing and transmission. By employing the extraordinary properties of quantum mechanical operations, such as superposition, entanglement, uncertainty, or complementarity, data may be encoded in the quantum states of matter or light and manipulated with unprecedented speed and efficiency.

The emerging technology of QCC has the potential to overcome many of the limits of the existing conventional technologies, offering not just incremental improvements but a fundamental breakthrough. QCC could revolutionize information processing, providing novel methods of securing, processing, storing, retrieving, transmitting, and even displaying information. Practical applications could include super-fast computers operating at the subatomic scale, and fully secure information transmission.

18.1.2 Sub-Sector Definitions

A. Quantum Computer Science

Quantum Logic Gates
A quantum logic gate, like a classical gate, is a very simple computing device that performs one elementary quantum operation, usually on one or two qubits. Any n-qubit gate can be constructed as a network from single-qubit and two-qubit gates. Quantum logic gates are able to create and perform operations on quantum superpositions.

Quantum Networks
Quantum networks are linked arrays of quantum logic gates. They are necessary for performing non-trivial quantum computation operations.

Error Correction (Fault Tolerance)
The fragile nature of the entanglements and superpositions employed by quantum information processors makes such systems extremely vulnerable to disturbance by noise. Quantum Error Correction employs networks of quantum gates and measurements to counteract the decoherence effects of noise in quantum systems. Such networks are themselves tolerant of errors, and are able to remove more noise than they introduce.

B. Experimental Techniques

Ion Traps
In an ion trap, a string of ions is confined by a combination of oscillating and static electric fields. Through manipulation with a laser beam, the ions can be made to change their atomic states or enter superposition states. These states can thus be used to encode qubits.

Nuclear Magnetic Resonance
In NMR the processor is a molecule. Each of the atomic nuclei within the molecule has distinct possible spin states, which can be used as the two states of a qubit. The molecule is placed within a magnetic field and the nuclear spin states (and thus the qubits) are manipulated through the application of oscillating magnetic fields.

Quantum Dots
Quantum Dots are semiconductor nano-aggregates having a very small number of electronic states. A change in the electronic configuration of the quantum dot, by the addition of an electron, of a hole or of an electron-hole pair, produces a

measurable change in the optical, electronic, or spin properties of the quantum dot, and so can be used to represent binary data.

Cavity QED
Inside an optical cavity the spatial and spectral distribution of the electromagnetic field is different from that in free-space. When a material system is introduced in a cavity whose dimensions are of the order of the wavelength of light, novel manifestations of the light-matter interaction may arise, permitting a complete control of the quantum features of the system. This can be exploited to code or process information at the quantum level.

Photon Communication
Photons, the individual particles of light, can already now be made with very high purity in a variety of states of individual photons and in entangled states. They can be sent over large distances, using either free-space transmission or optical fibers.

C. Technology

Quantum Input Output
Interacting with quantum objects presents fundamental difficulties, due to the introduction of a perturbation inherent to the observation act itself. Specific strategies, which can be termed as 'Quantum Input Output', have been elaborated in order to circumvent this limitation. Among these methods, the concept of Quantum Non Demolition measurement seeks a perfect preservation of the observable of interest during a measurement process.

Quantum Optical Networks for Communications
Quantum optical methods can be used to improve the systems performance of optical networks. Research topics on quantum optical networks includes low noise generation, amplification, detection and efficient information coding on optical quantum states. It also includes novel schemes for communication using specific quantum features of the electromagnetic field.

Delocalized Quantum Patterns
The quantum nature of light conditions the performance of optical techniques used in information processing, such as imaging and display, parallel processing and addressing of information, data storage, holography, high sensitivity detection and diagnostics etc.

Various nonlinear optical devices permit to exploit the quantum nature of light by generating spatial optical patterns with specific quantum properties, such as nonlocal quantum correlations between different points, entangled states of light, or locally reduced quantum fluctuations. They offer novel perspectives for information processing.

D. Applications

Quantum Cryptography
If the states of individual quantum systems are used for the transfer of conventional classical bits of information, eavesdroppers can always be detected. It is impossible to measure quantum systems in unknown states without disturbing them. Therefore, the integrity of the bits can be monitored; by using these bits as a crypto-key, the security of information transfer can be guaranteed.

Quantum Standards
Working with fundamental units of matter, such as single electrons and photons, allows a new level of accuracy to be determined with respect to absolute distances, time intervals etc.

Quantum Metrology
The techniques developed in tandem with the development of the quantum computer - like working with fundamental units of matter, such as single electrons, photons or photon-pairs - provide accurate and sensitive measurement techniques for determining quantities, like detector efficiency, source brightness, distances, time intervals etc.

Quantum Simulation
Simulation of evolving quantum systems involves processing the vast amounts of information involved in interference effects. In quantum systems the amount of relevant information increases exponentially with the number of particles involved. Predicting the way complex quantum systems interfere with each other therefore requires data processing that would take classical computers millions of years. Quantum computers should be able to carry out such simulations within reasonable periods of time.

Quantum Sensing
Entangled quantum systems are extremely susceptible to environment-induced decoherence. Hence, entanglement offers the highest possible sensitivity in the detection of changes in a surrounding environment.

18.1.3 Some Relevant Terms

Qubit
The qubit is a fundamental piece of information (i.e. a '0' or '1') written into a quantum system. As an example, the ground state of a hydrogen atom can represent '0', while its first excited state represents '1'. The qubit is analogous to the classical 'bit' of information, which can be represented, for example, by the presence or absence of a voltage on a capacitor plate.

Chapter 18: Categories and Definitions

Entanglement
Two or more quantum systems in superposition states can be linked together to form a system where each of the components loses its individuality. A disturbance of one component will affect all other components as long as they remain 'entangled' together.

Superposition
Quantum systems are able to simultaneously occupy different quantum states - this is known as a 'superposition' of states.

Coherence/Decoherence
Quantum systems in superposition or entangled states are said to be 'coherent'; this is a fragile state and easily disturbed by interaction with the environment. Decoherence occurs when a superposition or entanglement is disturbed and 'collapses', losing information to its environment.

18.2 The Disciplines that Contribute to Quantum Information Processing

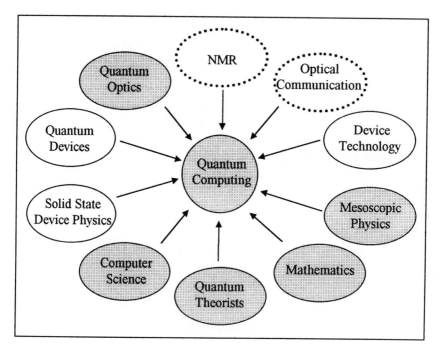

::: Relatively mature in terms of both science and technological development in the Quantum Information Processing context.

⬚ Considerable research work has been carried out in the Quantum Information Processing context, but still at the basic research stage, as far as QIP is concerned.

◯ Considerable basic research in the Quantum Information Processing context needs to be carried out before these fields can contribute to the subject.

Chapter 18: Categories and Definitions *143*

Disciplines Involved:

Quantum Computing:
All

Quantum Devices:
Quantum Physics, Solid State Physics, Semiconductor Physics

Quantum Optics:
Quantum Physics, Optical Physics, Optical Engineering

NMR:
Physics, Chemistry, Instrument Engineering

Optical Communications:
Optical Physics, Communications Engineering, Semiconductor Physics

Device Technology:
Solid State Physics, Semiconductor Engineering

Mesoscopic Physics:
Physics

Mathematics:
Quantum Mathematics, Statistical Mathematics, Chaos Theory Mathematics

Quantum Theorists:
Theoretical Physics

Computer Science:
Computer Science, Computer Engineering, Artificial Intelligence

Solid State Device Physics:
Solid State Physics

Appendix A
The Pathfinder Project

A.1 Pathfinder Activities

Following an application to the Long Term Research Project team in DG III/C of the European Commission, the Pathfinder Project was started in January 1998. This had originated in the British Computer Society, who were joined as partners by some of the leading experts in Europe in the subject of Quantum Information Processing. (For partners see A.2). The Project proposal study contract was approved, with a duration of one year and two clear aims:

1) To make a start on the task of bringing together the European workers in the broad field of Quantum Information Processing.

2) To draw up a report to the European Commission to describe the state of the field in the world and especially in Europe, giving recommendations for the future support of the subject in Europe.

The Pathfinder Partners held four meetings at appropriate centres in Europe to organise and monitor the work undertaken under the project. The Project produced a Newsletter which was circulated to some 1000 individuals and teams in Europe, and it is planned to circulate a further one later this year; a start was made on a database of 'Individuals and Organisations Active'; a taxonomy for the subject was drawn up, and a major conference was held, bringing together many, if not most, of the senior workers in the field in Europe.

A.1.1 Newsletters

In order to provide a vehicle for addressing and identifying individuals working and interested in the field in Europe, a 30-page Newsletter was drawn up. This contained an explanation of the Pathfinder project activities, an article reviewing recent developments in the field, book reviews, and publicity for the forthcoming Helsinki conference. It was circulated via the Pathfinder partners and from a central database to some 1000 individuals and research centres in Europe known to be interested in the field. It is expected that a further Newsletter will be prepared and distributed later in the year 1998, giving some information on the high-lights of the Helsinki Conference, reporting on some recent advances in the field, and outlining some of the material contained in this report. This will give an opportunity to draw attention to the conference publication, and to the research support opportunities that will become available for research and training support during 1999 under the Commissions Fifth Framework Programme.

A.1.2   Database of Active Individuals and Organizations

A start was made on preparing a database of individuals and organizations active in the broad QIP field in Europe. Partly due to the multi-disciplinary nature of the field, partly to the number of new individuals and organisations who are joining the field, this cannot be taken as a definitive list, but it does contain the contact details of some 250 individuals, together with information on the nature of their interests. This figure is not representative of the size of the active community in Europe, if only because it does not contain the names of all members of the major teams. But it does serve to demonstrate that the total number of workers in Europe in the broad field is still in the hundreds, not thousands as would be expected for a more mature subject. This work will need to be continued in some form in the future.

A.1.3   Taxonomy of the Subject

In order to classify the interests of the research workers and to provide some structure for considering the subject, a taxonomy of QIP was drawn up. (Annex II) This is a rapidly developing subject and this classification can only be considered as a starting point, which will evolve as the subject grows and stabilises.

A.1.4   Helsinki Conference

The Pathfinder Project team organised a three day conference in Helsinki at the end of September 1998, partly to bring the research community together to review the state of the subject, and partly to discuss what were felt to be the support needs for the field in Europe. The very full programme was deliberately designed to contain a considerable number of speakers, of whom 23 came from Europe (including one from Israel), 6 from the USA, and one from Japan and Australia. The conference was attended by over one hundred of the research workers from across Europe.

A.1.5 QIP Report

The main task of the Pathfinder project was seen as the preparation of a report to the Commission setting out the state of the art, and the considerations that will help the Commission to decide on their future approach to the subject. To facilitate the work, the project partners divided the field into four 'sectors', each under a leading figure in that research aspect, who prepared a report on the state of that sector, consulting other workers in the field as appropriate. These reports were circulated to those attending the conference and published on the World Wide Web. The reports were presented at Helsinki, and the last afternoon was devoted to discussion of them and of the needs of Europe. The Partners also attempted to establish what plans the National Funding Agencies, both in Europe and beyond, have for this field, though this proved difficult to carry out comprehensively in a rapidly developing field where many of the Agencies are in course of studying the situation.

A.2   Pathfinder Project Partners

Working Group:

Prof Anton Zeilinger, Institut für Experimentalphysik, Innsbruck University
Dr Izo Abram, CNET, Bagnol, Paris
Prof Giorgio Baccararani, DEIS, Universita di Bologna
Dr Guiseppe Castagnoli, Elsag-Bailey SpA, Genova, Italy
Dr Nicolas Cerf, Universite Libre de Brusselles
Prof Artur Ekert, Clarendon Laboratory, Oxford University
Prof Nicolas Gisin, Groupe de Physique Appliquée, Université de Genève
Dr Konstantinos Glinos, European Commission, DGIII, Brussels
Dr Miklos Santha, CNRS, Université Paris-Sud
Dr Antonella Karlson, StarLab, Riverland nv, Brussels
Dr Walter Van de Velde, StarLab, Riverland nv, Brussels
Dr Chiara Macchiavello, Pavia Universita (& Elsag Bailey), Pavia, Italy
Prof Dr Walter Schempp, Dept of Mathematik, Siegen University, Germany
Dr Peter Marcer, Cybernetics Machine Specialist Group, British Computer Society, London
Dr Brian Oakley (Chairman), British Computer Society, London
Mr Charles Ross (Hon Sec), European Institute of Quantum Computing, London
Dr Michael Brooks (Editor), Freelance Science Journalist

Appendix B
A Preliminary QIP Roadmap

	Research Goals	Applications
1987–1998	**BASIC QIP RESEARCH** • Quantum interference • Ion Traps • Cavity QED • Quantum Dots • NMR Schemes • Si:P • Shor's Algorithm • Quantum Cryptography • Grover's Algorithm • Single Photon Detector • Error Correction	**NOW** • Photon counters • Precision range finding • Optical time-domain reflectometry • Metrology • Secure optical communications
1998–2003	**NEAR TERM GOALS** • NMR to do 8 qubit QC • 3-5 photon entanglement • 3-5 qubit ion trap • Quantum dot gate • Silicon-based gate • Few qubit applications • New algorithms • Fault tolerant QC • Novel QC • Quantum repeaters • Quantum amplifiers • Detectors • Sources	**NEAR FUTURE** • Quantum cryptography demonstrators • General quantum communications • Random number generators • 2-photon metrology • DNA sequencing • Single molecule sensors • Simulation of quantum systems • Quantum gyroscopes
2003–2015	**MID TERM GOALS** • 10-20 qubit computation • Demonstrator systems • Novel algorithms • Macroscopic superposition • Bose condensate connections • Quasi-classical few particle gates and bits	**FAR FUTURE** • Quantum repeaters • Quantum amplifiers • Quantum computers (factorizing; non-structured information retrieval) • Molecular simulation
2015–?		

Subject Index

addends 79-80
Alice 12, 122-3, 87ff
atomic clock 46, 98, 110
attenuation 38

Babbage differential engine 17
Babbage, Charles 59
Bell inequality 46
Bell measurement 14
Bell states 11-12, 122
Bennett, Charles 5, 103
Bit Error Rate 35, 92
Bob 12, 44, 87ff, 122-3

cavity QED 6, 27, 30, 131
chloroform computer 29, 72
class P problems 18
Clear operation 74
computational complexity 18, 125
Controlled NOT gates 21-2, 27
Cory, David 46, 129
cytosine computer 29, 72

database searching 21, 97
decoherence 23, 42, 49-50, 129-30
decoherence times 28
defect tolerance 69
degenerate parametric amplifier 40
Deutsch, David 5, 23, 30, 103, 119
Deutsch's problem 29, 83-4

efficient computation 18
electron-nuclear double resonance (ENDOR) 54
ENIAC 66
entanglement swapping 14

entanglement, entangled pairs 3-4, 12, 41, 121, 125
EPR paradox 6, 119
exponential speed-up 20
fabrication plants 66
factorization 4, 18-19
fast/slow algorithms 18
Feher, George 54
Feynman, Richard 5, 103, 119, 128

genetic programming 109
Greenberger-Horne-Zeilinger (GHZ) states 122
Grover, Lov 5, 126
Grover's algorithm 5, 20, 126

Heisenberg uncertainty principle 44
Hilbert space 120, 128
house of cards effect 75, 98

information 9, 59-62, 120-1
initialisation 71
Intel 66
interference 23
ion traps 5, 24

Josephson junction 37

Larmor frequency 72
light, quantum nature of 34
lithography 56-7
Los Alamos 89

measurement 10, 38
metrology 46
microcavities 37

Moore's Law 66

NASA 109-11
no cloning theorem 120
noiseless branching 39
nonlinear optics 36
nonlinear refraction 36
NOT gates 22, 55
nuclear magnetic resonance (NMR) 6, 29, 71ff

observation 10
one-time pad cryptosystem 124
optical telecommunications 34

parallel computation 19
parametric downconversion 36, 46
parametric emission 39
Pauli principle 54
Penrose, Roger 104
phonon modes 2, 28
photodiode detection 34
photons 15, 24
plain adder 80
Planck, Max 5
public key 124

quantum cellular automata 113
quantum cryptography 14, 43-4, 87ff, 112, 123
quantum dense coding 12
quantum dots 6, 31
quantum error correction 23, 130-1
quantum fluctuations 34

quantum gyroscope 98, 110
quantum key distribution 87ff, 110, 112, 124
quantum logic gates 21
quantum networks 45
quantum noise 35
quantum non-demolition 38, 49
quantum repeaters 44
quantum simulators 45
quantum uncertainty 33

Rabi frequency 28
Readout 72
recoherence 23
RSA cryptosystem 17, 124

Schrödinger, Erwin 5, 12
Shor, Peter 5, 24
Shor's algorithm 17, 127
signal generation 37
Signal-to-Noise ratio (SNR) 35
solid state 53
spin-orbit coupling 54
squeezed states 37
Standard Quantum Limit (SQL) 35
stimulated emission 37
superconductors 31
superposition 3, 10, 120

teleportation 14, 123
Teramac 68
transistor 66
translating machines 97, 111
Turing machine 5, 119